Inventing the Cotton Gin

PUBLISHING FOR THE WORLD

125 Years

THE JOHNS HOPKINS UNIVERSITY PRESS

Johns Hopkins Studies in the History of Technology

Merritt Roe Smith, Series Editor

Inventing the Cotton Gin

Machine and Myth in Antebellum America

Angela Lakwete

The Johns Hopkins University Press
Baltimore and London

This book has been brought to publication with the generous assistance of the Robert Warren Endowment.

Johns Hopkins Paperbacks edition, 2005

9 8 7 6 5 4 3 2 1

The Johns Hopkins University Press
2715 North Charles Street
Baltimore, Maryland 21218-4363
www.press.jhu.edu

The Library of Congress has cataloged the hardcover edition of this book as follows:

Lakwete, Angela, 1949–
    Inventing the cotton gin : machine and myth in antebellum
America / Angela Lakwete.
        p. cm. — (Johns Hopkins studies in the history of
technology)
Includes bibliographical references and index.
    ISBN 0-8018-7394-0 (hardcover : alk. paper)
    1. Cotton gins and ginning—United States—History—19th
century. 2. Inventions—United States—History—19th century.
I. Title. II. Series.
TS1585.L35 2003
609.73′09′034—dc21   2002156776

ISBN 0-8018-8272-9 (pbk.: alk. paper)

A catalog record for this book is available from the British
Library.

Contents

Preface

The cotton gin animates the American imagination in unique ways. It evokes no images of antique machinery or fluffy fiber but rather scenes of victimized slaves and battlefield dead. It provokes the suspicion that had Eli Whitney never invented the gin, United States history would have been somehow different. Yet cotton gins existed for centuries before Whitney invented his gin in 1794. Nineteenth-century scholars overlooked them as well as gins made by southern—and northern—mechanics, in order to create a history meant to chasten some southerners and demean others. Using the gin as evidence, they read failure back from the Civil War into the choices that southerners made from the American Revolution, tracing the steps that led them to Appomattox.

The scholars compounded the idea of southern failure by ascribing agency not to individuals but to a machine. Writing in 1893, historian James Ford Rhodes acknowledged Whitney's gin as "one of the very first of those splendid mechanical inventions, which are justly our boast and pride," yet he blamed it for "riveting more strongly than ever the fetters of the slave" since it "prevented the peaceful abolition of slavery." It facilitated the production of cotton, which, he believed, required "slave labor." Since "cotton fostered slavery; [and] slavery was the cause of the war between the states," the gin was doubly culpable.* It "fostered" both slavery and cotton, and its invention, according to Rhodes, was the first step toward disunion. Late twentieth-century historians have replaced the belief that technology determines historical events with explanations that emphasize the reciprocal influence of individuals and technology on sociopolitical change. These revisions have overturned many once-cherished beliefs but have scarcely affected standard textbook interpretations of the gin.

*James Ford Rhodes, *History of the United States from the Compromise of 1850* (London: Macmillan and Co., 1893), 25–27.

The lineage of interpretations like this can be traced to an 1832 biography of Eli Whitney written by Yale College alumnus Denison Olmsted. Benjamin Silliman commissioned and published it in his prestigious *American Journal of Science and Arts*. Olmsted portrayed Eli Whitney as a mechanically adept adolescent whose early life prefigured his invention of both the cotton gin and interchangeable parts manufacturing. The cotton gin section begins in 1792, when Whitney visited the Mulberry Grove plantation near Savannah, Georgia, on his way to a tutoring position in South Carolina. Olmsted included statements about the ineffectiveness of roller gins, used from the colonial period. He asserted that enslaved Africans in the South used no technology other than their fingers to remove cotton fiber from seed before Whitney invented the gin. Whitney's involvement in interchangeable parts manufacturing began in 1798, when he secured a contract from the federal government to manufacture muskets. He used the advanced funds to pay debts incurred in defending his gin patent. Olmsted did not conclude that Whitney achieved interchangeability. Instead, Olmsted wrote, Whitney revolutionized arms manufacture by increasing and simplifying the number of stages in the manufacturing process, allowing a manufacturer to hire "inexperienced workmen" at a lower cost. But he insisted that Whitney invented the first cotton gin. On the last page of the biography, the editor added a drawing of Whitney's tombstone with the epitaph that begins, "Eli Whitney, The inventor of the Cotton Gin," graphically ending the argument.*

This study reopens the discussion. It explores the history of the gin as an aspect of global history and a facet of southern industrial development. It begins with cotton, the commodity the gin was invented to process, then examines gin invention and innovation in Asia and Africa from the first to the seventeenth century, when the British colonizers introduced an Asian hand-cranked roller gin to the Americas. First indentured British and later enslaved Africans built and used foot-powered models to process the cotton they grew for export on mainland North America and in the Caribbean. Starting in the middle of the eighteenth century, cotton production increased in response to expanding textile production in Great Britain. Colonial mechanics correspondingly built water-, wind-, and animal-powered models, all of which were types of roller gins. In 1794 Eli Whitney patented not the first but a new type of gin. As cotton

---

*Denison Olmsted, "Memoir of the Life of Eli Whitney, Esq.," *American Journal of Science and Arts* 21 (Jan. 1832): 208, 214.

growers and users debated the utility of his wire-toothed gin, southern mechanics transformed it into the saw gin. Industry adoption of the saw gin in the mid-1820s coincided with the social and economic upheavals that shaped antebellum America. Southern gin makers maintained their dominance, patenting and manufacturing saw and roller gins along with McCarthy and cylinder gins, two new gin types invented in 1840. Cotton planters encouraged the southern gin makers, northern gin makers competed with them, and Confederate ordnance officials contracted them. The innovative antebellum southern gin industry belies constructions of failure read back from 1865. Instead, it forces a reconciliation of an industrializing, modernizing, and slave labor–based South.

Acknowledgments

Many dedicated librarians, archivists, museum curators, agricultural engineers, gin managers, as well as professional and amateur historians have contributed to this work, and I thank them all for their tireless efforts. The book began as a dissertation in the Hagley Program for the History of Industrialization in the Department of History at the University of Delaware, and I express my gratitude to the program and to the department. Historian of science and technology George Basalla, as my academic advisor, facilitated a difficult career transition by teaching me how to use "things in history." He also chaired the dissertation committee and guided me through the process with gentle resolve. David F. Allmendinger, historian of early national and antebellum America and dissertation committee member, pushed me out of what unknowingly had been a shallow intellectual life into the chasm of primary documents and textual analysis—from which I hope never to emerge. Economist and historian Farley Grubb and historian of technology Arwen Mohun, both dissertation committee members, helped me develop analytical tools and historical perspectives that deepened my understanding of the historical process. Historians Anne Boylan and Wunyabari Maloba taught, advised, and encouraged me, both providing models of professionalism in their commitment to innovative instruction and research.

I thankfully acknowledge funding that supported preparation of the book manuscript from the Smithsonian Institution National Museum of American History, the Huggins-Quarles Award of the Organization of American Historians, a Grants-in-Aid from the Early American Industries Association, and the Chancellor's Postdoctoral Fellowship at the University of California, San Diego.

Of the many individuals who helped me track down sources, I must thank particularly Dr. Marion J. Rice, Cavett Taff, Clare Sheridan, and Suzanne Bell. Dr. Rice is a University of Georgia professor emeritus and former director of the Clinton Historical Society in Clinton, Georgia. As director and member, he oversaw the restoration and documentation of a 40-saw saw gin attributed to Samuel Griswold, an antebellum gin maker. He graciously shared his extensive knowledge of textile and gin history through images, documentation, and citations but also through discussion and correspondence. In his capacity as curator of exhibits at the Old Capitol Museum of Mississippi History in Jackson, Cavett Taff supplied me with and directed me to sources on the Old Southwest from its colonial, territorial, and early statehood periods, including gins in museum collections. His knowledge of the Delta, the Trace, and antebellum planter personalities infused life into historical fact. Mr. Taff continues to invigorate regional history as a freelance consultant and exhibit designer. As director of the Osborne Library of the American Textile History Museum in Lowell, Massachusetts, Clare Sheridan was of inestimable assistance. She combed her considerable collections and found documentation on gins and gin makers, including infringement suits, broadsides, and catalogues, particularly relating to the defunct cylinder gin but also on roller and saw gins. Suzanne Bell was a 1998 Smithsonian Institution intern who worked tirelessly reading through dozens of reels of microfilmed census data for evidence of gin makers. Without the enthusiastic support of these professionals, research would have been less interesting and the work less complete.

The manuscript benefited from close, critical readings conducted by several exceptional individuals. Dr. Basalla read the earliest versions but also later revisions. He brought to bear on a conceptually and structurally rough work his knowledge of technology and artifactual analysis as well as his experience as a writer and a series editor. He persevered, parrying my often-defensive responses and urging me to think deeper about my sources and my interpretations. Dr. Pete Daniel, Smithsonian Institution curator and historian of the twentieth-century South, read early versions of the work as well as related conference papers. Dr. Daniel also allowed me access to the Smithsonian's extensive collection of Asian and American cotton gins and related pictorial and documentary files. Mr. Ed Hughs, engineer and research director of the USDA, Agricultural Research Station, Southwest Cotton Ginning Research Laboratory in Mesilla Park, New Mexico, was involved in the project at an early stage and read several early and later versions. At the laboratory, which includes a library and archive, he reshaped my academic knowledge into a more practical understanding of gin mechanics and the cotton market. Ms. Nobuko Kajitani, head textile conservator at the Metropolitan Museum of Art in New York City, nurtured my love of fiber and fabric during the nearly ten years I worked in her laboratory, and beyond. She applied a lifetime of making, researching, and analyzing fabrics to a close reading of a manuscript version. Dr. Anthony G. Carey, historian of the Old South at Auburn University, Alabama, closely read a late version of the manuscript. I must also thank Dr. Carolyn C. Cooper, who read and commented on an early draft of the chapter on Eli Whitney. An anonymous press reviewer read an early and a late version of the manuscript. I am profoundly grateful for the unsparingly critical yet supportive comments on both readings. They forced me to examine tone and attitude as well as content and organization, and improved the final version fundamentally. I must also thank series editor Dr. Merritt Roe Smith and acquisitions editor Dr. Robert J. Brugger for their long-standing interest in the project and their patience during the revision process. Last, I thank my family and friends, neighbors and colleagues for all their support.

Inventing the Cotton Gin

CHAPTER ONE

Cotton and the Gin to 1600

Cotton gins were used to remove cotton fiber from the seed since the first century of the Common Era. The earliest gin was made from a single roller and a hard, flat surface. Ginners in Asia, Africa, and North America used it to batch process fuzzy-seed, short-staple cotton. Sometime between the twelfth and fourteenth centuries, gins built with two rollers replaced it in commercial Indian and Chinese markets, although the single-roller gin persisted in the domestic sphere. The evidence for these changes is scarce and confusing. It suggests that while several types of gins may have been used in the Byzantine, Muslim, and Mongol Empires, by the sixteenth century an Indian-style two-roller gin had become the dominant gin in the Mediterranean cotton trade. In the seventeenth century, British merchants' interest in the cotton trade collided with their pursuit of the spice trade in the Indian Ocean and of colonies in the Americas. At the same time, British industrialists responded to the domestic craze for colorful Indian cotton prints by initiating industrialization based on the production of cheap substitutes for the expensive imports. Hand- and foot-powered spinning wheels and looms facilitated industrial expansion; cotton—processed by hand- and foot-powered roller gins—fueled it.

The history of the gin begins with the cotton plant, a tropical perennial that grows in a swath around the globe between 47° north and 35° south latitude. Neolithic farmers domesticated cotton about ten thousand years ago for its seed or fiber, or both. Linnaeus grouped the plants into the genus *Gossypium* (Order Malvales). Today, botanists recognize fifty species, four of which are cultivated for their seed and fiber. Two are the Old World species *G. arboreum* and *G. herbaceum;* two are the New World species *G. hirsutum* and *G. barbadense.* Three of the four—the two Old World species and one of the New World species, *G. hirsutum*—share a seed characteristic that is relevant to the problem of fiber removal.

Cotton fiber is a single cell of cellulose that grows out of the seed as the plant matures and fruits. The fruit of the cotton plant is a compartmentalized boll filled with fiber-covered seeds. When the fiber is removed from the seed of the three species mentioned above, a fuzzy covering remains. Botanists once believed that the fuzz was evidence of poor ginning—that it was the broken ends of fibers damaged in the process of removal. They now believe that the fuzz is a chemically distinct extrusion called linters. Lint is the ginned fiber used in fabric production, whereas linters are short, coarse, unmeduled, and unspinnable fibers that grow out of the seed five to ten days after the lint. On the basis of the fuzzy-seed characteristic, *G. arboreum, herbaceum,* and *hirsutum* can be grouped into one category. They are commercially known as *short-staple, green seed,* and *Upland* cotton, although fiber length, seed color, and viable locale vary. However, they are all "fuzzy-seed cotton" and were all ginned with types of roller gins.[1]

The New World species *G. barbadense,* on the other hand, grows in a limited habitat. It produces a smooth seed and was ginned literally by hand before 1600. The species is indigenous only to South America, although it diffused to southern Central America and limited regions of the Caribbean. It is not native to Barbados, despite its Linnaean classification. According to the nineteenth-century English botanist George Watt, Linnaeus had no sample of *G. barbadense* in his herbarium. He relied on the notes and drawings of Leonard Plukenet, a seventeenth-century English naturalist, and, according to Watt, misunderstood them.[2] British settlers had introduced cotton to Barbados in the 1630s after settling the island in 1627. They cultivated cotton from seed purchased from Dutch merchants in Guiana or Brazil.[3] The Dutch appropriated the seed from one of the world's oldest cotton cultures. The people of the Amazon River Basin had cultivated *G. barbadense* for millennia and produced knotted and netted fabrics that date to 2500±110 B.C.E.[4] Rather than invent a machine to process their cotton, they modified the plant.

The seed of *G. barbadense* has a unique morphology. It extrudes no linters and the fiber attachment to the seed is weak, relative to the fuzzy-seed species. Fiber removal thus results in a smooth, bald seed. In all cotton species, the boll contains an average of fifty fiber-covered seeds. In most varieties, the seeds are discrete elements from which the fiber is removed. This was not the case for the Brazilian variety of *G. barbadense*. Ancient Americans modified the cotton plant to produce seeds that fused into one kidney-shaped mass.[5] Rather than remove fiber from many small seeds, they processed fewer, larger seeds. That, in conjunction with the weak attachment of the fiber to the seed, allowed ginners to remove fiber quickly and safely with only fingers and thumb. Native South Americans used biology not technology to transform finger ginning into an efficient batch process.

Cotton producers who cultivated the fuzzy-seed species (*G. herbaceum, G. arboreum,* and *G. hirsutum*) faced the problem of removing the fragile fiber that adhered to tiny seeds. Removal was necessary before farmers could use the seed or artisans, the fiber. They solved the problem by inventing the gin. The word *gin* signifies ingenuity and skill, and is appropriately applied to the first machine invented for fiber removal. The device was an ingenious Neolithic construction, easy to make but hard to use. Like the inclined plane or the lever, the gin transmitted and redirected the ginner's force and required a high degree of skill. Effective ginners developed the dexterity needed to process the seed cotton without breaking the fiber or cracking the seed.

The first gin was also the most persistent. Scholars documented its existence in the fifth century C.E., and it was still being used in the late twentieth century. They located it on three continents, in Asia, Africa, and the American Southwest, where Native Americans cultivated *G. hirsutum*. Whether it spread from one culture or originated independently in many cultures, it was an archetype that established the principles that would govern gin development for centuries. Its inventors may have been the cotton producers, the fiber users, or specialized artisans peripheral to the cotton culture. The anonymous inventors understood mechanics and appreciated the need to preserve the integrity of the seed and fiber in the process of fiber removal. They may have first used finger and thumb to remove fiber, but outside of South and Central America, this was only a transitional technique.

Referred to as an instrument and a utensil, the first gin was a two-part machine consisting of a narrow roller and a flat base. It paired an active with a passive element; together they allowed the ginner to batch process seed cotton (fig. 1.1). The roller was the active element and the more specialized of the two components. Gin makers fashioned it from an iron rod and varied its length from ten to twenty-four inches. It overhung the passive ginning surface by a hand span on either end. It was thicker in the center and tapered to flat or rounded ends in order to compensate for its tendency to spring, or bend away from the surface in use. The base varied in size and shape. Gin makers textured it so that it could stabilize the seed cotton without impeding fiber separation. Asian gin makers favored rectangular wooden bases, while West Africans preferred oval stone surfaces.

The ginner used the roller and the ginning base together as the single-roller gin. Typically she placed the base on the ground in front of her and layered it with seed cotton, the mass of cotton and seed removed from the ripened boll of the plant. She leaned over the base and grasped the overhanging ends of the roller. Ginning occurred when she rolled the roller over the seed cotton, pinching the seed out of the lint. The posture allowed the ginner to apply the weight of her upper body but to vary it as needed. The ginner then pushed the seed away from the roller; the lint accumulated behind it.

Since the single-roller gin so closely resembles the saddle-quern (or metate), a machine used to grind grain, one would expect it to date from the Neolithic Period, like the quern. Both have active and passive elements. The active element of the quern is a thick roller, usually made of stone. The passive element is a flat, upward-sloping base typically made of a hard, textured stone, on which the miller places the grain to be ground. Like the ginner, the miller puts the base on the ground and kneels in front of it to add the weight of her upper body to the roller as it cracks the grain. The only significant difference between the two machines is the diameter of the roller. The diameter of the ginning roller cannot exceed about five-eighths of an inch. Anything thicker crushes the cotton seeds within the wider angle it forms with the base, as the larger quern roller does with the grain. Except for this minor difference, the two machines are identical.

Archaeologists' oversight may explain the absence of evidence that would locate the single-roller gin in prehistory. That the rollers of extant gins are made of iron does not preclude the possibility that the machine predated the Iron Age. The roller could have been made of stone like the narrow-diameter rollers of some Neolithic Chinese saddle-querns. It could also have been made

Fig. 1.1. Single-Roller Gin, Mende, West Africa, twentieth century. The two-part single-roller gin consists of an iron roller (approx. 12″ long) and a flat base (approx. 5″ wide × 4″ high × 8–10″ deep). The thickened center of the roller compensates for springing. The spinner grasps the tapered ends, which overhang the base. Brigitte Menzel, *Textilien aus West Afrika* (Berlin: Museum für Völkerkunde, 1972), plates 15, 16. Courtesy, Staatliche Museen zu Berlin, Preußischer Kulturbesitz, Ethnologisches Museum, Berlin.

of hard wood, since its purpose was to separate rather than crush the seed. Without associated seed fragments or fiber, however, rollers or bases—whether made of wood or stone—would be difficult to identify as parts of a gin. Even at Iron Age sites, correct attribution would be strained by the similarity of the gin parts to other tools. An iron roller might be catalogued as a spindle, the base as a spindle support rather than as a ginning surface. On the other hand, archaeologists may have recorded no single-roller gins because they found none. The finger ginning of fuzzy-seed cotton may have persisted into the Common Era.

A fifth-century Buddhist painting in the Ajanta Caves in the western state of Maharashtra, India, constitutes the earliest evidence of a single-roller gin. The painting depicts a woman moving a large-diameter roller over a flat surface. At first glance, she appears to be using a saddle-quern, and scholars alternately described her as grinding spices and cosmetics. Linguist Dieter Schlingloff reinterpreted the image in light of other Buddhist drawings and Jain texts. He concluded that the woman was using the single-roller gin, not the saddle-quern. Two other women in the scene provided the key to understanding the image. In Schlingloff's reading, the center figure is grasping a mass of seed cotton in her hand and is poised to hand it to the ginner. The woman to her right is holding a stylized cotton bow needed to fluff the fiber after ginning. The competing interpretations of the painting highlight the similarity of the saddle-quern and single-roller gin, but by contextualizing the main figure, Schlingloff offered a convincing identification of the earliest documented cotton gin. It was this gin, Schlingloff argued, and not the familiar *churkha*, or roller gin, that was the original gin of India.[6]

From the fifth to the nineteenth century, Indian women used the single-roller gin to process fuzzy-seed cotton. It persisted in India after the introduction of other types of roller gins and steam-powered saw gins. It was a traditional technology but one that was not entirely static. Gin makers introduced a foot-powered model that took advantage of a woman's lower body strength. Rather than kneel on the ground, the ginner sat on a stool and placed the gin under her feet. The numerous references to both hand- and foot-operated models that appeared in British and U.S. technical literature at the end of the nineteenth century may have emanated from a few British sources. Engineer and amateur historian John Forbes Watson (who published as J. Forbes Watson) originated one stream. In 1866 he republished an excerpt from *A Descriptive and Historical Account of the Cotton Manufactures of Dacca in Bengal* by "a former Resident of Dacca." Originally published in 1851, it described a woman placing a small amount of seed cotton on a "smooth flat board . . . then rolling an iron pin backwards and forwards upon it with the hands, in such a manner as to separate the fibres without crushing the seeds." In the 1870s Watson traveled to Dharwar and saw a foot-operated model. It had been used in India, along with the *churkha*, "for time immemorial," he wrote, remarking that both models preserved the staple but crushed some seeds.[7] These and other nineteenth-century descriptions of single-roller gins verify that the gin persisted and that gin makers modified it incrementally over time.

Evidence situates the hand-powered single-roller gin elsewhere in Asia and western Africa; it was also found in the American Southwest, but only theory can suggest how it arrived there. It seems certain that the Native Americans who cultivated *G. hirsutum* invented the gin independently, since there is no evidence that European colonists transferred the technology to the Americas.[8] The diffusionist theory may explain its appearance in China, central Asia, and western Africa. Botanists identify *G. herbaceum* var. *africanum*, a variety indigenous to southwest Africa, as the ancestor of linted cottons in the Old and New World, but historians believe that the first Old World cotton complex originated in the Indus River valley.[9] In support of the initial Indian complex, scholars cite the fabrics of Mohenjo-daro, which date to between 2300 and 1760 B.C.E. (±115) and the written accounts of Indian cotton by, among others, the Greek historian Herodotus in the fourth century B.C.E. and Aristobulus, Alexander's historian, in the third century B.C.E. Descriptions of tree cotton in Arabia preserved by Theophrastus, a pupil of Aristotle, and Androsthenes, an admiral in Alexander's army, are used as evidence of its diffusion.[10]

Archaeological evidence dates cotton production in the upper Nile River valley to forty-five hundred years ago, suggesting its persistence through the dynastic Egyptian and Nubian periods, although linen predominated as the fiber of choice.[11] The Roman poet Virgil documented the "cotton that whitens the Ethiopian glades" in his *Poetic Treatise on the Art of Farming,* written in 29 B.C.E.[12] Archaeologists were able to verify evidence of cotton production and textile manufacture in Meroë in the fourth century C.E. Historians continue to debate whether it was an indigenous industry or an extension of the Indian complex.[13] Except for Aristobulus writing in the second century B.C.E., none of the sources on India or Africa mention ginning.[14] Strabo quoted Aristobulus in his *Geography* dating from the first century C.E., describing Indian "wool-bearing trees," the flower of which "contains a seed, and that when this is removed the rest is combed like wool."[15] Aristobulus confirmed seed removal but not the use of the gin.

Not until the thirteenth century did references to the single-roller gin appear in documentary sources. Its use, and documentation, is associated with the expansion of cotton production throughout the Muslim Empire. From their roots in Mecca, a diverse urban center near the Red Sea coast of the Arabian Peninsula, the followers of Islam diffused a syncretist belief system after the death of its founder, the Prophet Muhammad, in 632 C.E. Islam combined elements of Judaism, Christianity, and Arabic culture into an inclusive monotheism without the complex doctrines that wracked seventh-century Christianity. Muslims took advantage of the internal weaknesses of the Byzantine and Sasanian Empires to capture territory and trade routes, creating an empire that linked three continents. At its height, it encompassed eastern, northern, and parts of western Africa, southern Spain, and much of western Asia, including Turkey, the Levant, and central Asia. Muslim merchants and diplomats created and maintained trade linkages to China and Southeast Asia. The Muslim synthesis diffused ideas and commodities throughout the Afro-Eurasian ecumene that changed as they were acculturated. Islam also imposed change. Muslims, the followers of Islam, were required to cover their bodies completely, revealing the face, hands, and feet only. Elite Muslims wore silk and linen, but most people fulfilled the obligation with garments of cotton. Islam created new demand for low-cost, lightweight cotton fabric and was the means for intensifying cotton cultivation where it existed and diffusing it into new regions, perhaps along with the Indian single-roller gin.

Although Chinese farmers cultivated cotton before the seventh century, Islam may have provided an expansionist impulse. Archaeologists recovered cotton fabrics that date from the sixth century in Kaoch'ang (Turfan) and Lop Nor, both in the Xinjiang region of northwestern China, but scholars disagree on their source. Some believe that the fabrics verify documentary sources on indigenous production and manufacture, but others argue that they were introduced as trade goods.[16] The process of diffusion from the west may have begun in the seventh century, when Muslims visited the Tang court and, with the approval of emperor Yung-Wei, built a *masjid,* the Muslim place of communal worship, in the capital city Ch'ang-an. By the Song Dynasty (960–1279 C.E.), Muslims dominated commerce but their involvement in agricultural production was not recorded.

In 1273, just before the fall of the Song Dynasty to the Mongols, a scholar published an agricultural manual that documented single-roller gin use.[17] He described the roller as an "iron staff . . . two feet long and one finger thick, the two ends gradually tapering, like a rolling-pin." The base was made of "pear-wood, three feet long, five inches broad, and two inches thick." The ginner placed "the floss . . . on the board; the iron staff is then 'rolled' to drive out the seeds, and one gets 'pure' cotton."[18] As described, the components appear mismatched. The roller overhung the base by more than two hand spans and the base was longer than would seem practical. The author may have exaggerated its proportions to convey structure and mechanics, but he nevertheless described a gin that was similar to nineteenth-century Indian gins. The similarity, however, confuses vectors of diffusion. It is plausible that the Chinese gin mentioned in the thirteenth-century document had diffused west to India and influenced gin design there. But it was at this time that cotton first made its appearance in paper and began competing with silk as a textile fiber.[19] Inferential evidence, therefore, points to the diffusion of a cotton complex from India stimulated by the Muslim Empire.

In the Middle East, expansion of the Muslim Empire coincided with the introduction of cotton culture and the use of the gin. Although King Sennacherib of Assyria may have cultivated *G. arboreum* in his irrigated gardens about 700

Fig. 1.2. Single-Roller Ginner, Iran, late fifteenth century. West African and Iranian ginners would sit at the gin, not stand as shown. His roller would have resembled the tapered African roller (see fig. 1.1); both conformed to thirteenth-century Chinese descriptions. By permission of the British Library, Acc. No. PL03990.

B.C.E., the plant is not believed to be indigenous to the Middle East; wool was the fiber of Mesopotamia.[20] Cotton flourished after Muslims introduced it to the region, defined as including the area surrounding the Caspian Sea as well as the Levant.[21] Not until the fifteenth century, however, was the gin documented in use. An Iranian manuscript painter identified ginning as one of the four "Stages in the Manufacture of Cotton Cloth." He illustrated a man standing at a gin that seems to float in the air (fig. 1.2).[22] To the painter, the man's costume was as important as his occupation, so he employed an artistic convention to display both. Undoubtedly the ginner would have sat at the gin, like thirteenth-century Chinese and nineteenth-century Indian ginners.

In western Africa, Muslim emissaries and merchants documented cotton as a luxury fabric but never recorded the details of cotton production or processing that they either found there or introduced. Muslims began their conquest of western Africa in the eighth century, when the ancient state of Ghana controlled the salt mines in the Sahara Desert north of the Niger River and the gold fields in the tropical forest south of the river valley. Cotton textiles were peripheral to Muslims' interest in these lucrative trades, yet in the eleventh century Al-Bakri wrote of the "lengths of fine cotton" that the citizens of Sila, a city-state on the northern bank of the Senegal River, used for currency.[23] The Syrian geographer Al-Idrisi confirmed Al-Bakri's observations in the twelfth century. He described the wealthy people of Sila and Takrur, a state whose rulers adopted Islam about 1030, wearing "garments of cotton and mantles," observing that commoners dressed in wool.[24] Cotton was a luxury in the thirteenth century, when Al-Dimashqi described nomads south of Sijilmasa wearing "skins except for the few who wear cotton."[25] He believed that they bought the fabrics from merchants in Kawkaw, a city on the Niger River.

Cotton may have been produced in the Niger and Senegal River valleys before the Muslim conquest, but scholars are more confident that it was cultivated on a large scale after conquest. If Arab Muslims introduced cotton, it would seem probable that they also would have introduced the single-roller gin, if not in the tenth century, when the citizens of Sila and Takrur had converted to Islam, then in the thirteenth, when Islam became the state religion of Mali. Mansa Musa, the state's famous emissary, organized an elaborate pilgrimage to Mecca in 1324. Although his primary purpose was to fulfill a religious obligation, the king spread the wealth and knowledge of Mali. The pilgrimage might have initiated the transfer of the Indian cotton complex to western Africa or the expansion of an indigenous industry.

Europeans observed cotton production and manufacture in tropical western Africa in the seventeenth century, but not until the nineteenth century did they record the use of the single-roller gin. William Finch, a British merchant, visited Sierra Leone in 1607 and reported that inhabitants planted "Cotton, called Innumma, whereof with a Spindle they make a good Threed, and weave it artificially, making cloth thereof a quarter broad, to make coverings for their members, and being sewed together, Jackets and Breeches."[26] Interested in process and product, Finch nevertheless did not mention ginning. The cotton complex he saw persisted through the ravages of the eighteenth-century Atlantic slave trade. This complex was disrupted—but not destroyed—by the influx of mass-produced British trade goods. In 1846, French anthropologist Anne Raffenel described it, recording the use of the single-roller gin. Conducting fieldwork in Mali, Raffenel watched female ginners prepare the seed cotton by drying it. They then "forcibly pressed it between the stone and a small metal rod (made by the ironworkers) held in the hand."[27] Similar accounts followed, like those by Watson in 1866 and 1879, with anthropologists eventually photographing the gin and single-roller ginners in the twentieth century.[28]

The spotty evidence of the single-roller gin confirms its use in the American Southwest and in the Indus, Yangtze, and Niger River valley societies, yet its low outturn doomed it in expanding cotton markets and led to the introduction of a more efficient gin. In 1893 a French official measured the outturn of the gin in then colonial West Africa at one-half kilogram or about one pound of lint per day, an amount that can serve to approximate earlier figures.[29] The gin had become a bottleneck in Indian and Chinese cotton economies perhaps as early as the twelfth century. In response, gin makers introduced the roller gin, a gin with two rollers powered by either one or two cranks. One variation of the roller gin diffused throughout India, Southeast Asia, and the Levant. In the seventeenth century, it traveled from there to the Americas on French and British merchant ships. Two variations, both invented in China in the fourteenth century, did not transfer across cultures but may have influenced gin development in American colonies in the mid–eighteenth century.

The roller gin was harder to make but easier to use than the single-roller gin, and represented another stage in the mechanization of the pinch principle of fiber removal. This gin was composed of two narrow-diameter rollers that rotated simultaneously but in opposite directions. The gin maker wedged the rollers, each typically no more than twelve inches long, into a vertical frame. Indian gin makers added a worm gear that allowed the ginner to use one crank to turn both rollers. As they rotated, the rollers pulled in the fibers between them, pinching the seed off. Since they were wedged tightly together, the rollers held back the seed, which dropped off after the fiber was removed. Fiber removal was no longer dependent on the skillful application of the single roller. Instead it occurred automatically with the turning of the rollers. One late nineteenth-century source cited an average outturn of five pounds per day, five times that of the single-roller gin.[30] In expanding medieval cotton economies, the faster roller gin pushed the single-roller gin into marginal markets and became the dominant commercial gin.

One piece of evidence for this change is a twelfth-century mention of a ginning technology that has been interpreted as the roller gin. The reference was documentation, if not of the shift in ginning technologies, then of the successful appropriation of the techniques of Muslim cotton complex by Venetian merchant-manufacturers. In the twelfth century, Venetian entrepreneurs began making inexpensive copies of the luxury cotton imports manufactured in state-controlled Muslim *tiraz* workshops, underselling them in European markets. From the first imports of cotton fiber in 1125, the industry grew rapidly. Venetian textile makers imported fiber from cotton-producing regions throughout the Mediterranean Basin. They procured lower-quality stocks from North Africa and southern Spain as well as Apulia, Malta, Sicily, Alexandria, Antioch, and Armenia. Producers in the three Syrian centers of Hama, Aleppo, and Acre supplied higher-quality staple. Cyprus was another major center. All of them supplied both the Muslim and Venetian cotton textile industries. Operating on Muslim privileges, Venetian merchants coordinated storage and outfitted cogs to ship the lint to Venice. The enormous quantities shipped suggest that the worm gear–driven, hand-cranked roller gin had displaced the single-roller gin, but the evidence is inconclusive.[31]

Reference to a *buttarello*—"cotton gin" in the Apulian dialect—situates a gin in an Apulian dowry contract in 1110. Maureen Mazzaoui, a historian of the medieval Italian cotton industry, has speculated that the term referred to the roller gin and that it might have come from Sicily.[32] Documentation from Syria followed—but not until the late fifteenth century. This source was an edict issued by the governor general of the Province of Hama in 1489. He ordered "that the tax should be abolished, which was levied on the cotton merchants at Hama, a tax which was levied on the utensils of the cotton ginners."[33] Literally carved in stone in the Great Masjid at Hama, the edict provides clues to the status and organization of the Levantine cotton industry. Significantly, it identified merchants, not cotton producers, as owners of gins. That alone might argue for the use of the larger, more complex roller gin. But the word "utensils" suggests the roller and base of the single-roller gin. If "utensils" were used to gin cotton in a major Syrian cotton-exporting center in the late fifteenth century, then the term *buttarello* may have referred to the single-roller gin.

The earliest documented roller gin was not the familiar Indian *churkha* but a two-person, gearless Chinese gin. A Chinese scholar described and illustrated it in *Nung shu*, an agricultural manual published in 1313. He depicted the two ginning rollers wedged into an upright rectangular frame and attached to a bench. The gin maker had made the upper roller out of iron and the lower roller out of wood, a choice that reflected the composition of the single-roller gin. In the drawing, both rollers extend through the upright member of the frame, forming part of the cranks that the ginners used to turn them.[34] Two men operated the machine: one of them turned a crank and supplied the seed cotton, and the other turned the second crank in the opposite direction and removed the lint. The author of *Nung shu* mentioned the single-roller gin but recorded that the ginners of the Yuan Dynasty used this two-person machine.

Cotton cloth was currency in Mongol China; in such a market, a gin that could be operated by one ginner replaced the productive but labor-intensive two-person model. The one-person gin was mentioned in a 1360 manual but pictured in *T'ien-Kung K'ai-Wu* (A volume on the creations of nature and man), published in 1637.[35] The gin maker had attached a crank to one roller and a treadle to the other. The illustrator depicted the male ginner effortlessly cranking one roller with his left hand, treadling the other with his right foot, and supplying the gin with seed cotton with his right hand (fig. 1.3). Later drawings show the gin head attached to a table and the ginner seated on a stool.[36] Structurally, the one-person, hand-and-foot–powered roller gin was a logical progression from the two-person gin. It added no new mechanics while halving the labor and increasing ginner productivity.[37] Gin makers later added a flywheel to the roller that the ginner treadled, a modification that may have increased outturn but also risked ginner fatigue or injury.[38]

The worm gear–driven Indian roller gin, known to many as the *charkha* (although *belna* is a more accurate term), became more widely used than either Chinese gin. The gear allowed a single ginner to turn both rollers in opposite directions with a hand crank with one hand and supply the rollers with seed cotton with the other (fig. 1.4). This machine has persisted to the present virtually unchanged. Gin makers introduced a version with a narrow iron upper roller and a large wooden lower roller, allowing them to build a wider gin while maintaining the proper angle between ginning surfaces to avoid crushing seeds. Other innovations suggested underlying problems with the worm-gear mechanism. It was a technical marvel but difficult to carve and sensitive to the environment. Only a slight deviation in the width of the screw made cranking difficult; warping rendered the gin useless. The iron roller and gear section may have partially solved the problem. Indian gin makers also introduced a two-person direct-drive gin that was similar to the Chinese two-person gin while also meeting the gender requirements of Indian work traditions. Rather than straddle a bench like Chinese male ginners, the Indian woman sat on the ground and supplied seed cotton as she cranked one roller. The second ginner was a man who sat on a stool and used a stick to turn a flywheel attached to the

棉赶

烘火

Fig. 1.3. The One-Person Roller Gin and Ginner, China, 1673. The ginner uses his left hand to crank one roller and his right hand to supply the gin while his right foot treadles the opposite roller. Sung Ying-hsing, *T'ien-Kung K'ai-Wu: Chinese Technology in the Seventeenth Century,* trans. E-Tu Zen Sun and Shiou-chuan Sun, (University Park: Pennsylvania State University Press, 1966), 61. Courtesy, Dover Publications, Inc., Mineola, N.Y.

opposite roller.[39] It is unknown whether the ginners coordinated roller revolutions or if synchronized revolutions were necessary. The gin demonstrated that Indian gin makers experimented with different configurations in order to overcome mechanical problems and increase efficiency.[40]

The fragmentary evidence of early gin history raises more questions than it answers. One question about the choice of gins is particularly relevant to the

Fig. 1.4. Hand-Roller Gin with Worm Gear, India, twentieth century (approx. 16½″ wide × 14″ high). Note the handle, or crank, pivoting off the lower roller and the worm gear on the opposite side. U.S. gin makers would call the flat piece of wood underneath the lower roller a "guard." It prevents lapping and adds structural integrity. Smithsonian Institution, Photo No. 67865.

American experience. From about the twelfth to the seventeenth century, the Indian roller gin and the Chinese one-person gin coexisted within the Muslim Empire. Although there are no outturn figures with which to compare the two machines, it would appear that the Chinese gin would have allowed for greater outturn and ginner productivity. It exploited the stronger leg muscles of male ginners rather than the weaker arm muscles of female ginners. The machine itself was easier to make and more adaptable. Chinese woodcuts show it strapped to benches and tables as well as standing alone. Yet it was the less efficient Indian roller gin that diffused into Southeast Asia and western Asia, while the Chinese gin did not breach its cultural boundaries. The question of why merchants did not introduce the more efficient Chinese gin into the large eastern Mediterranean cotton economy must be asked even though it cannot be satisfactorily answered.

Answers may lie in opportunities forgone in Mongol China. Overcoming the Southern Song Dynasty in 1279, Mongols from the northern steppe created an empire that spanned two continents. At its height, it stretched from China to eastern Europe, incorporating the principalities of Russia and much of the Anatolian peninsula, Mesopotamia, and central and Southeast Asia, including India. In 1260, a resolute Mamluk Empire stopped expansion into Africa, and an indomitable Japan prevented similar expansion in the East. Seduced by silk textiles brocaded with gold yarn, the Mongols resettled defeated Iranian textile artisans in China, where they learned from and taught Chinese weavers.[41] Conqueror and conquered alike became agents of transfer, traveling the "Silk Road" with cotton as well as silk, gold, and spices. At the beginning of the dynasty, Mongols were fierce persecutors of Muslims, but by its end they had adopted the faith. Until their defeat in 1368, they were a bridge linking China to the Muslim world. The dislocations as well as the linkages created during the Mongol reign were opportunities for technology transfer. Silk manufacturing technologies traversed the Silk Road but cotton-processing technologies apparently did not. There is no evidence of the worm gear–driven Indian roller gin in China or of the Chinese one-person roller gin in India, the Levant, or elsewhere in the Muslim Empire. Instead, the evidence again points to the diffusion of "utensils," a word suggestive of the single-roller gin.

Nevertheless, it was the Indian roller gin equipped with the worm gear that diffused east to Southeast Asia, west to the Levant, and ultimately to the Americas on British merchant ships in the seventeenth century. Understanding how the British became participants in the cotton trade and agents of transfer requires a brief review of the familiar history of the spice trade and the initial stages of British industrialization. Bronze Age Britons did cultivate flax, but it was wool that defined the British textile industry from the Roman Period. Cotton was so rare in Britain that in the Middle Ages it was mythologized as the product of a "vegetable lamb," a fantastic creature that resembled a sheep but grew on a stalk like a plant.[42] But the British were not entirely unfamiliar with cotton. They used the fiber to stuff quilts and yarn to make lamp wicks, an import that the English Wardrobe Act of 1212 regulated. The British also purchased imported fustian, a fabric made of linen and cotton that was a staple of the Venetian textile industry from the twelfth century. As it declined, the south German textile industry expanded and carried on fustian manufacture from the late fourteenth to the early seventeenth century.[43] It was then that British "drapers" began making import substitutes in Lancashire. Middlemen supplied them with linen produced domestically or in Ireland, but the cotton came from British "Turkie Merchants" stationed in the Levant.[44]

British involvement in the cotton trade of the Levant began in the fifteenth century as a consequence of the competition among European powers over access to the luxuries of the East. It was sparked by Muslim imperialism, to which the Crusades had responded in the twelfth and thirteenth centuries. Competition intensified when Ottoman Muslims conquered Constantinople, the capital of the beleaguered Byzantine Empire, in 1453. Emboldened Muslims annexed Bosnia and Herzegovina ten years later, triggering an appeal for a crusade by Pope Pius II. Christians ignored the call but not the sentiment. The Reconquest of the Iberian Peninsula and the search for land and sea routes that bypassed the Muslim-controlled Levant were two processes that had been underway before 1453 but accelerated in the face of Ottoman aggression. Hugging the coastline, the Portuguese cleared the southern tip of Africa and sailed east into the Indian Ocean in 1498. Christopher Columbus, an Italian mariner from Genoa, persuaded the Spanish monarchy that a westward tack across the Atlantic Ocean would lead to the Indian Ocean in less time. He set sail in August 1492 and landed in the Caribbean Sea thirty-three days later.

The 1494 Treaty of Tordesillas confined Spain to the Western Hemisphere, where the monarchy exploited Columbus's blunder while other European powers converged in the Indian Ocean, intent on seizing control of the high-stakes spice trade from Muslim merchants. The most coveted spices were nutmeg, mace, and pepper, which were used as medicinal and recreational drugs in Europe. Sailing agile and armed carracks, the Portuguese mounted successful assaults along Mughal India's southwestern Malabar Coast in the 1490s. In 1510 they captured Goa and later Gujarat to the north, both important cotton textile centers. Malacca, the gateway to the Indonesian archipelago, fell in 1511. By 1535 the Portuguese controlled the key ports in the Indian Ocean and the trade routes to the Spice Islands.

In the late sixteenth century, the British launched a two-pronged challenge to Venetian and Ottoman dominance of the trade in the eastern Mediterranean Sea and to Portuguese interests in the Indian Ocean. Queen Elizabeth chartered a series of joint-stock trading companies mandated to circumvent and undermine British trading rivals. The Muscovy Company received the first charter in 1553 committing it to discover a northern passage to the East.

The Venice and Turkey Companies were chartered as trading companies. The Turkey Company received its charter in 1581, the Venice Company in 1583, both for seven years. They were to exchange British-made goods, particularly woolen fabrics, for eastern commodities, especially spices, currants for wine, and also silk fiber and fabrics and "cotton wool [fiber]" and yarn. That Turkey merchants engaged in the cotton trade was borne out by the few papers that remain. A document dated only to the Elizabethan period identifies English merchants trading "cloth and kerseys" among other articles for "Indigo, raw silk, spices, drugs, currants . . . cotton wool and yarn." Another document dated to 1590 confirms that merchants upheld the terms of their charters exchanging British-made manufactures for "cotton wool and yarn, cotton cloth—blue and white" and other luxury goods.[45]

When the Turkey and Venice Companies' charters expired, they merged into "The Governor and Company of Merchants of the Levant" and did business as the Levant Company. Anxious to maintain trade and political alliances with the Ottoman Empire, Queen Elizabeth approved the charter in 1592. Merchants secured concessions from Ottoman rulers and established embassies and factories (actually warehouses) in the capital city of Istanbul, formerly Constantinople, and in Smyrna, Aleppo, Alexandria, and other cities of the Levant where they continued to exchange British for eastern goods. Their success in trade and diplomacy led King James I to renew the Company's charter when it expired in 1605. The principals renamed the company "The Governor and Company of Merchants of England Trading to the Levant Seas" and it became an even more important conduit for English woolen textiles. Parliament affirmed its monopoly in the Levant by issuing an ordinance in 1643 that congratulated the company for providing the means "whereby the poor people are sett at worke," in their "venting of kerseys, sages, perpetuanas, . . . above 20,000 broadclothes per annum, besides other commodities."[46] Exporting woolen fabrics, Levant Company merchants imported increasing amounts of cotton fiber and yarn, secured in Smyrna and Cyprus, that they used in the expanding Lancashire fustian industry.[47]

On the last day of 1600, Queen Elizabeth chartered the British East India Company whose mandate infringed on the Levant Company's monopoly in the silk and cotton trades; its directors, however, chose to explore markets in the Indian Ocean rather than in the Mediterranean Sea. Shortly after receiving the charter, they dispatched an envoy to the Mughal Indian court of Jalal ad-Din Akba. Diplomatic efforts bore fruit when his son, Emperor Jahangir, extended a privilege to the company allowing it to build a factory in Surat, a city on India's northwest coast, in 1612. By then the Dutch had undermined Portuguese hegemony in the spice trade. Dutch merchants chartered the Dutch East India Company in 1602 after an earlier mission to the Indian Ocean had returned a wealth of spices to Amsterdam. They returned to wrest the trade from the Portuguese, using existing tensions between them and the local rulers to their advantage. Like the Portuguese and the English, the Dutch were interested not in settlement but in trade, building a factory in Surat in 1616 joining the British.

Control of the spice trade, not the textile trade, was the goal of the British East India Company, but the two trades converged in the early seventeenth century. British merchants were at first successful in doing business in the Indian Ocean by exchanging English woolens and fustians and Indian cotton textiles and other manufactured goods, although competition eventually forced them to use specie.[48] When the British attempted to claim territory in order to control access to the region's resources, they encountered determined opposition. The Dutch trounced them at Amboyna (also spelled *Amboina*, today *Ambon*) in 1623 and went on to defeat the Portuguese at Malacca in 1641. The Treaty of Breda ended the war in 1667; by its terms, England renounced its remaining claims in Southeast Asia. The British East India Company had in the meantime expanded its investment in the cotton trade. In 1664 alone, it shipped a quarter million lengths of printed cottons from India to England.[49] The fabrics filled a rising demand for colorful, colorfast, all-cotton fabrics and sparked the domestic production of inexpensive substitutes for the expensive imports. This process accelerated an import substitution–driven industrialization that had begun in the fustian industry. This was an industrialization fueled by hand- and foot-powered spinning wheels, looms, and cotton gins.[50]

By the seventeenth century, the trade in cotton fiber, yarn, and fabrics had moved from the periphery to the forefront of British mercantile policy. The policy had been initiated to circumvent the Muslim and Catholic middleman but it had pushed Protestant British merchants into closer contact with them. Richard Hakluyt fanned the flames of religious intolerance in his 1584 *Discourse of Western Planting,* but he also articulated the logic of Atlantic colonization, in the hopes of generating enthusiasm among potential investors. Colonies in the Americas, he wrote, would "yelde unto us all the commodities of Europe, Affrica, and Asia, as far as wee were wonte to travel, and supply the wantes of all our decayed trades."[51] Hakluyt imagined British citizens producing the currants, citrus fruits, silk, and cotton that merchants were buying from Muslim and Venetian middlemen. With the settlement of Virginia Colony in 1607, his dream materialized. British merchants transferred cotton to the Americas, along with other Asian and African "exotics"—confident that they could overtake the Old World trades.

CHAPTER TWO

The Roller Gin in the Americas, 1607–1790

With the transfer of the Asian cotton complex to British North America in 1607, the history of the cotton gin in America began. Following the dictates of mercantilism, British colonists located cotton production and processing in the American colonies and cotton textile manufacture in England. Indentured servants cultivated fuzzy-seed Old World cotton and used roller gins to process it. Mainland producers floundered in the competitive cotton trade, however, and sidelined cotton by the end of the century in favor of tobacco, rice, and indigo. Caribbean producers stayed in the market after they introduced cotton in the 1620s. By the turn of the century, they supplied more fiber to British textile makers than the historic Levantine producers. Mainland North American producers reentered the trade in the 1770s after British inventors mechanized cotton spinning. Mechanization increased demand for cotton fiber and pushed downstream change in gin design. Gin makers in the Americas responded conservatively with variations on the roller model. They invented a fully foot-powered gin by mid-century, introduced the barrel gin in the 1770s, and in 1788 Joseph Eve invented a self-feeding gin that could be powered by water, wind, or animals. When cotton prices soared in the 1780s, statesmen exhorted planters to expand cotton production. Their response was to use a variety of roller gins to process the crops. Their growing dominance in the global cotton trade at the end of the eighteenth century redeemed the missteps at the beginning of the seventeenth.

On the tenth of April, 1606, James I, King of England, Scotland, France, and Ireland, and Defender of the Faith, chartered the joint-stock Virginia Company of London, dividing the "sondrie Knightes, gentlemen, merchanntes, and other adventurers" into the Virginia and Plymouth Companies. He enjoined both to settle and develop designated sections of the east coast of North America from the 34th to the 45th parallel but selected the Virginia Company as the "Firste Colonie" to set sail.[1] Three ships set out from London on December 20, 1606, and landed at Chesapeake Bay on April 26, 1607. Christopher Newport, captain of the *Susan Constant* (also called the *Sarah Constant*), directed the settlers to plant the seedlings and seeds they brought with them, before he left to explore the newly named James River. One month later, a "Gentleman" reported back to London investors that the "West-Indy plants of oranges and cotton-trees thrive well; likewise the potatoes, pumpions, and mellions."[2] He distinguished between the Old World orange and cotton crops, possibly grown from seedlings, and three New World crops, which he may have found growing in the gardens of the Algonquian-speaking people of the region. He added that colonists had also "carefully sowne . . . garden-seeds," which also "prosper well." The Gentleman may have exaggerated New World fecundity to reassure armchair investors like Richard Hakluyt, a member of the "Firste Colonie." But his 1607 report revealed that Europeans had already transferred Old World cotton, *Gossypium arboreum,* to the Atlantic economy and that English colonists had planted it in Virginia Colony. He did not mention a gin but he also neglected to mention the tools that the "laborers" used to plant and tend other crops. Since the Company intended colonists to produce for export, it can be inferred that they shipped with the colonists the equipment they needed to be successful.[3]

However, a decade of "killing time" and "starving time" passed before settlers exported any commodity, extracted or produced, from the colony. Contradicting the Gentleman's sanguine account, Captains Newport, John Smith, and Gabriel Archer reported fatal confrontations with Powhatan and his men—all of which portended tragedy.[4] Not until 1617 did colonists begin to turn a profit—but with the unlikely New World crop tobacco. To King James's dismay, his subjects had adopted the "vile custom of tobacco taking." Their fondness for what he called a "filthy novelty" in a 1604 polemic created a demand, which the Virginia Colony ironically yet conveniently met.[5] Investors temporarily forgot Richard Hakluyt's entreaty that colonists reproduce Old World exotics in the New World colonies to eliminate dependence on Spanish and Ottoman Muslim merchants. Instead, they pursued the extraction of iron and naval stores; the dyestuffs madder, woad, and fustic; and produced "silke grasse" for mattress stuffing and hemp and flax for cordage and yarn, among a host of products indigenous to the colony.

The *Records of the Virginia Company* demonstrate the company's continued interest in cotton production through the early 1620s but also their failure to make it profitable. Cotton appears eighth on a 1620 list of fifty-four "Commodities Growing and to be Had in Virginia." It was highly valued at "eyght pence the pound" and "had in abundance," according to a June "Declaration."[6] That the company expected indentured servants to grow it is revealed in a September 1620 indenture granting servants land for "plantinge of olyves, sewing or plantinge of cotton woll Anyseed, Wormseed and the like."[7] In a 1622 "Note" recapitulating the progress of the Colony during the previous year, a settler reported that "The Plants of Cotton-wooll trees that came out of the West Indies, prosper exceeding well, and the Cotton-wooll-seeds from the Mogols Countrey come up, and grow: Samples of it they have sent; and this commodity they hope this yeere to bring to a good perfection and quantity."[8] Planters had not relied on seedlings from the Caribbean, the document reveals. They imported cotton seed from Mughal India, "Mogols Countrey." But this last reference is telling. Old World cotton grew in the New, but colonists had not produced much of it. Having introduced it in 1607, colonists sent only a sample to London investors in 1622 and hoped to ship more in the future.

In the Caribbean, English colonists grew cotton alongside tobacco and sugar, succeeding at shipping merchantable fiber probably ginned with the roller gin, although they too failed to mention any details of processing. In 1624 they claimed the uninhabited island of Barbados as Crown property and peopled it with indentured servants three years later. They grew cotton along with tobacco from seeds and plants that English merchants had procured from the Dutch in Surinam, receiving high prices for both crops in European markets.[9] Profits plummeted over the next decade and planters subsequently replaced both crops with sugar cane after 1640. The crop change paralleled the shift from indentured British to enslaved African labor. After seizing Jamaica from the Dutch in 1655, the English cultivated cotton and tobacco along with spices and condiments. When Jamaican planters shifted to "sugar and slaves," two planters followed an opposite course, substituting cotton for sugar in 1684.[10] Caribbean cotton producers maintained production levels even during periods of low prices. The Levant Company, operating at cross purposes, increased its cotton exports from Smyrna and Cyprus during the mid– to late seventeenth century, possibly depressing prices and hurting English producers in the Caribbean.[11]

Successful production of cotton in Barbados inspired the Lords Proprietors of the Carolina colony to introduce it after settlement in 1663. By 1690, cultivation was so widespread that the proprietors used it as currency, accepting rents in seed cotton. But nine years later they suddenly ceased production. According to Edward Randolph, colonists turned to naval stores and rice because cotton could no longer "answer their expectations."[12] Tobacco had met Virginians' expectations by the 1620s; naval stores, and later rice and indigo, Carolinians'. The Carolinians, however, continued to produce cotton for domestic use. A resolution passed by the House of Commons in 1746 that allowed "clean [that is, ginned] Cotton and dressed Hemp Be made a Tender in Law" suggested that substantial quantities were grown.[13] The House may have revived the 1690 ordinance allowing cotton to be used as currency or legalized a customary practice. Apart from its fiscal connotation, the resolution confirmed John H. Lawson's 1709 report of Carolina women making "a great deal of Cloath of their own Cotton, Wool and Flax."[14]

While mainland producers marginalized cotton, Caribbean cotton producers overtook the cotton growers of the Levant as suppliers to Britain's expanding fustian and cotton textile industries over the course of the seventeenth century. Once used only for candlewick and padding, cotton had become a major industrial input. In 1621 fustian makers presented a petition to Parliament identifying the sources of their raw materials, suggesting the size of the industry, and revealing their notions of civic responsibility. It read in part,

> about twenty years past diverse people in this Kingdome, but chiefly in the Countie of Lancaster, have found out the trade of making of other Fustians, made of a kind of Bombast or Downe, being a fruit of the earth growing upon little shrubs or bushes, brought into this Kingdome by the Turkie Merchants, from Smyrna, Cyprus, Acra and Sydon, but commonly called Cotton Wooll; and also of Lynnen yarne most part brought out of Scotland, and othersome made in England, and no part of the same Fustians of any Wooll at all. . . . There is at least 40

thousand peeces of Fustian of this kind yeerely made in England . . . and thousands of poore people set on working of these Fustians.[15]

The fustian makers looked to "The Turkie Merchants" of the Levant Company for lint and to the Scottish merchants for linen yarn. The industry, a decentralized putting-out system of production, employed "thousands" of indigents, they wrote, who spun the short cotton fiber on the same hand-operated spinning wheels they used to spin wool. Beneficiaries of the declining south German fustian and cotton industries—themselves heirs to the collapsed Venetian cotton industry—the Lancashire fustian makers used foreign raw materials to supply local and distant markets with domestically manufactured fabric.

Fragmentary evidence suggests the pace at which the fustian industry expanded and the speed with which Caribbean cotton producers overtook their Levantine competitors. In 1626 the London partnership of the brothers George and Humphrey Chetham handled "Ciprus wool," that is, cotton fiber imported from Cyprus, probably by the Levant Company. The brothers sent twenty "packes" to Manchester, shipped another 36.5 packs "downe," and conducted several retail transactions totaling 1,094 pounds. Marking all but three packs as weighing 240 pounds, they handled approximately 14,654 pounds of fiber. By the end of the century, cotton factors like the Chethams also handled fiber from India exported by the British East India Company and "West Indian cotton" shipped by independent merchants. The Custom House recorded these details. For the year 1697 to 1698, merchants imported 390,419 pounds of fiber from the Levant, 849,598 from the Caribbean, 25,629 from India and Africa, and 202 pounds from an unknown source, totaling 1,265,848 pounds.[16] At the beginning of the century, cotton imports were negligible; at its close, they exceeded one million pounds a year. Sixty-seven percent of the total for one documented year came from Caribbean colonies.

In the early eighteenth century, the cotton gin attracted the interest not of merchants or colonists but intellectuals, who documented machines and practices, some of which may have dated to the previous century. The details of production and processing fascinated Enlightenment philosophers, and they included close descriptions of cotton and cotton gins in their encyclopedias. The English *philosophe* Ephraim Chambers published the first in his 1728 *Cyclopaedia.* Cotton, he wrote, was "a sort of Wool, or rather Flax encompassing the Seed of a Tree of the same Name." It grew "in several Places of the *Levant,* and of the *East* and *West-Indies,* especially in the *Antilles.*" Making a "very

considerable Article of Commerce," cotton must be avoided as a wound dressing, he cautioned. Leeuwenhoek himself had examined it under the micro-
scope, Chambers reported, and found that "its Fibres [had] each two flat
Sides" that cut the flesh, aggravating injuries.[17]

Of the gin, Chambers wrote, "The Seed of the *Cotton* being mix'd, in the
Fruit, together with the *Cotton* it self, they have invented little Machines,
which being play'd by the Motion of a Wheel, the *Cotton* falls on one side, and
the Seed on the other; and thus they are separated."[18] The entry located the
roller gin in the French Caribbean in the early eighteenth century. Although it
was "little," Chambers distinguished it by the using the term "Machine." How
it operated is not entirely clear from the description. The "Wheel" he mentioned could signify several different mechanisms. It could refer to the rollers,
to the worm gear of an Indian-style gin, or to a larger roller or drum, which in
later gins turned the rollers. The description of the separation of the fiber and
seed conforms to others and confirms the use of a roller gin. Colonists may
have made the gin themselves, as Chambers thought, but it is more likely,
from other evidence, that it was shipped to the Antilles from the Levant.

Competing with the British in Asia, the French continued the rivalry in the
fledgling cotton trade in the Americas. Shortly after Jean-Baptiste le Moyne,
Sieur de Bienville, founded the city of New Orleans in 1718, he established
cotton cultivation. In 1726 he wrote that cotton "grows in all the places of the
[Louisiana] colony where it has been planted." Colonists grew a perennial
plant, the roots of which did "not die during the winter" but sprouted new
branches in the spring. "It will some day be the occupation and the wealth of
the small settlers and of the poor people," Bienville accurately predicted. Yet
planters devoted only 1.4 percent of their land to cotton in 1731.[19] If Chambers is to be trusted, French colonists in the Caribbean possessed an effective
gin, but mainland French apparently had none or could not use those they
had. Without a gin, colonists understandably limited cotton production and
urged colonial officials to encourage the invention of an effective machine.

A shipwright referred to only as Isaac invented the first in 1733. He built a
"mill through which, by means of two steel drums joined to each other and
turned by a double wheel, the cotton passes very easily and the seed remains
. . . Two negroes will turn the wheels and two little negroes who will feed the
mill will be able to gin fifteen to twenty pounds of it per day," the Commissary General Edmé Gatien Salmon explained to Count Maurepas.[20] This large
roller gin promised to alleviate the problem of ginning. Salmon received per-

mission from Maurepas to reward Isaac with "a small gratuity" of "five or six hundred livres." Enthusiasm was short lived, however, and planters ultimately declared the gin a failure without explaining the reasons. Others continued experiments through the 1730s and early 1740s with similar results.[21]

A new commissary general, Mezy Le Normant, applied fresh insights in 1746 to what had remained a problem for mainland French cotton producers. He addressed the organic along with the mechanical, modifying the plant, presumably through seed selection, in the hopes of producing a strain more easily ginned with the roller gin. He then ordered that scale drawings be made of two gins in his possession. Medical doctor and botanist Louis Prat had built one; Le Normant called the other simply "Ancien Moulin Fait à la Louisiane" (old gin made in Louisiana). The drawings conveyed a European interpretation of the Asian roller gin bereft of the Indian worm gear. Both French gin makers westernized the gin by substituting toothed gears, but Prat went further. He separated the crank mechanism from the rollers and attached it to a large toothed gear that intermeshed with the roller gears from a position below them (fig. 2.1).

The change in gear types may have reshaped the gin into a more familiar artifact, but it presented a formidable challenge to the gin maker and ginner. If the gears were carved of wood, the small teeth would have intermeshed only with difficulty and broken easily. In the process, they would exert forces against the rollers that would push them apart, springing them. Casting iron gears to the necessary size and tolerance would have daunted a pattern maker and founder, particularly in the colonial setting. Prat tried to overcome the problem by adding thumbscrews at the top of the frame. Presumably the ginner could use them to keep the rollers together. The result, however, would be a gin that required even more force to turn. With good intentions, Le Normant sent the drawings to Paris, requesting that a gin be built from them. In response, he received a shipment of two gins in 1750, made on a conventional plan for cotton growers in Guyana on the northeastern coast of South America. The sender, who included no description of the gins, wrote diplomatically that he hoped they would "be useful."[22]

French colonists in the Caribbean did not share the ginning problems of mainlanders, nor, it seemed, did they share their gins. Prompted by Chambers' 1728 encyclopedia, the French *philosophe* Denis Diderot published his monumental *Encyclopédie* between 1751 and 1772. In one of the eleven volumes of engravings, published in 1762, he printed a plate depicting "an island

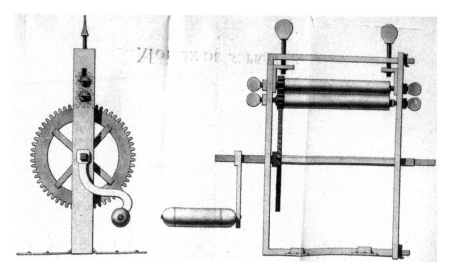

Fig. 2.1. "Moulin du Sr. Prat," 1746. Prat, a French colonist in Louisiana, replaced the Indian worm gear with toothed gears. Note the two thumbscrews on the top of the frame used to adjust the space between the rollers. Courtesy, C.A.O.M., Aix-en-Provence (Archives nationales, France), F3/86 (folio 230).

in America where cotton is grown." It depicts an idealized cotton plantation where enslaved African men picked cotton, a woman ginned it, and men packed it. Another plate includes a larger drawing of the roller gin in the plantation scene and a smaller, Indian-type roller gin.[23] The *Encyclopédie* conclusively documents the transfer of the Indian roller gin to the Americas, but it also documented a new type of roller gin that the woman in the scene used.

Two overstuffed bags of cotton fiber dominate the foreground of the "island in America" plate, but the ginner, tucked behind cotton baggers and overlooking a bay, draws attention as the only machine user. She sat at a large cotton gin, which she treadled with her feet. The treadle was attached to two large flywheels that were themselves attached to rollers. A larger drawing of the foot gin in an adjacent plate clarifies how the treadle was attached to the flywheels, but not how the treadle was affixed to the rollers. The ginner rested her forearms on a ledge built into the gin just below the rollers. It held the seed cotton, which she pushed into the rollers. As she ginned, the fiber fell into a basket placed on the ground behind the rollers; the seeds popped off in front of them.

The foot-powered roller gin represented a major advance in gin design that may have had Asian as well as American roots. The entry on the gin in the *Encyclopédie* suggested that both it and the hand-roller gin were familiar machines. It distinguished them only by the source of power, the foot-powered gin being *"[un] moulin à pie[d]"* and the hand-powered gin as *"[le] moulin à main."*[24] Within the lengthy description of the mechanics of the foot gin was no suggestion that it was new or remarkable, perhaps because the French were not the only users of the gin, or the earliest. In 1747 the Englishman James Marion petitioned the General Assembly of South Carolina for the "sole Privilege of making & licensing" a gin that he claimed could process eighty pounds of "rough Seed Cotton" in a day.[25] The assemblymen did not read a description of the gin into the minutes, but the outturn of twenty-six pounds, approximately one third the weight of the seed cotton, matches that of later foot-gin figures. Significantly, Marion's petition documents interest in—if not the use of—the roller gin on "rough" or fuzzy-seed cotton in the North American colonies. The *Encyclopédie* entry may have described Marion's gin, or one like it. The engravings may have illustrated either or both French and English foot gins. Taken together, the evidence of Chambers, Marion, and Diderot situates roller gins in the Caribbean and the North American mainland by the first half of the eighteenth century.

The juxtaposition of the hand and foot gins in Diderot's *Encyclopédie* validated colonial references to imports of hand gins from the Levant. It also blunted the impact of the foot gin. Positioned next to the hand gin, the foot gin seemed a logical, and unremarkable, development. Yet no gin maker had made the conceptual leap involved. Chinese gin makers had approached the idea in their one-person, hand-and-foot roller gin. They introduced it in the fourteenth century, but it had not diffused west through either the Muslim or Mongol Empires. Instead it had remained confined within the borders of China and had become the workhorse of China's extensive cotton industry. Europeans in the Americas were newcomers to cotton production and the gin. As such, they would appear to be unlikely inventors of the new foot gin. Were any of the enslaved involved in cotton culture in Africa, they would have used the single-roller gin, not the Indian-type roller gin. But it was the foot gin that became the workhorse of the American cotton industry in the eighteenth-century and persisted through the first half of the nineteenth. Questions about its origin or the source of the idea that inspired it are relevant not only to the history of the gin but to an understanding of the vectors of technology transfer and of the pace of technological change in the Atlantic economy.

An answer to its origins may lie in Ming and Manchu (Ch'ing Dynasty) China, where British East India merchants undoubtedly saw the one-person, hand-and-foot gin. The 1623 massacre of Amboyna, in which Dutch soldiers tortured and killed at least sixteen British East India Company factors, ultimately drove the British out of the spice trade. The Company retreated to Surat on the west coast of India, where it had established a factory in 1612 and from which it deepened its involvement in the cotton trade. It also expanded into the China trade, wedging its way into Canton, a trading center that Portuguese had penetrated in 1517. In the waning months of the British civil war, London directors advised Surat factors to formalize a trade agreement with the city of Canton in 1648. But China itself was engulfed in war, as Manchus fought and ultimately conquered the Ming Dynasty. Not until 1699 did British merchants gain a foothold in the trade; Cantonese officials did not confer privileges until 1704.[26] From their factory on Taiwan, British East India Company merchants attempted to fulfill the terms of their charter, trading British woolens and other British-manufactured goods for Chinese porcelains, teas, and cottons.

The details of the trade during the first half of the eighteenth century confirm that British merchants interacted with Cantonese merchants who dealt in cotton fiber as well as fabric. An established cotton industry, including producers and manufacturers, surrounded Canton, a city on the southeastern coast of China. It was also the entrepôt of a regional cotton textile trade. A measure of British interest was a 1736 shipment of 10,374 "pieces" of "nankeen," a cotton fabric woven in Nanjing, the capital of the Ming Dynasty. It was prized for being more colorfast than cloth woven in Canton. Other shipments recorded in the same account included 2,560 pieces of "Nankeen cloth" and 2,010 pieces of "Canton cloth," in addition to tea, sago, and "chinaware." The merchants also traded ginned cotton. They purchased bales of cotton in Bombay—250 in one 1739 shipment—and sold the fiber in parceled amounts in Canton.[27]

The trade in cotton fiber, if not in cloth, could have placed English or French merchants in or near a ginnery. It is not unreasonable to imagine that they could have sent a sketch or a description of the Chinese gin to company directors or even shipped a gin to colonists in the Americas through the company. Yet, even if they had—and there is no evidence that they did—colonists in the Americas would still have had to make the conceptual leap from the Chinese hand-and-foot gin to the foot gin illustrated in Diderot's *Encyclopédie*. It appeared that gin makers in the Americas eliminated the hand crank and attached both rollers to a treadle. It was an important evolutionary step, one that Chinese gin makers had not taken.

The foot gin further mechanized fiber removal on the pinch principle. It required less energy to operate because it allowed the ginner to use either or both feet to treadle and both hands to supply seed cotton. Theoretically able to treadle faster and work longer, the ginner increased her productivity, producing more fiber with less effort than with the hand roller gin. Of uncertain origin, then, but likely an American invention, the foot gin eliminated the hand-cranked roller gin, the first to be used in the emerging Atlantic cotton trade, and became the dominant gin.

The advantage, if any, the foot gin gave American over Levantine cotton producers is difficult to determine. There are no records of the number of foot gins in use in the Atlantic economy or of hand gins in the Mediterranean economy and no records of the average outturn of either. Additionally, the rising Caribbean imports against declining Levantine imports in British Custom House accounts are no measure of relative productivity gains. All things being equal, British merchants would have followed the dictates of mercantilism and purchased colonial over foreign products. Statistical data confirm that cotton production expanded in the Caribbean and that, over the course of the century, British merchants imported more fiber from there than from Levantine sources. In 1750, for example, British Customs Officials recorded 1,085,383 pounds of fiber imported from the Levant and 1,125,936 from the West Indies. Five years later they recorded 1,572,485 from the Levant and 2,188,362 from the West Indies. Ten years later, in 1765, the ratio had changed only slightly in favor of West Indian producers, but by 1770 imports from the Levant dropped to 764,600 while those from the West Indies increased to 2,838,816. In 1780 British merchants imported only 338,743 pounds of cotton from the Levant and 4,166,122 from West Indian growers.[28] By this time, the center of the cotton trade, once situated in the Mediterranean economy and Asia, had shifted to the Atlantic economy. Englishmen might have remembered Hakluyt's 1584 *Discourse* and believed his dream fulfilled.

The mechanization of yarn spinning in Britain in the eighteenth century drove the expansion of cotton production and the invention of more efficient gins in the British North American colonies. The mechanization of spinning

had begun centuries earlier when an anonymous artisan invented the spinning wheel by mounting a spindle on a frame and belting its whorl, or flywheel, to a larger wheel. As the spinner turned the wheel, the spindle rotated
rapidly, imparting twist to cotton or wool fibers as she pulled out the mass of
fiber to spin a yarn. Winding on was a separate operation. The spinner briefly
stopped the wheel and reversed it to back the yarn off the tip of the spindle,
then rotated it normally to wind the yarn she had spun onto the base of the
spindle. She then repeated the process, spinning, attenuating, and winding.
Historians speculate that Indian artisans invented the spinning wheel sometime between 500 and 1000 C.E. and that it traveled within the Muslim Empire
to Europe, where artisans built a larger model. European women stood at the
great, or Jersey, wheel and used it to spin both cotton and wool fiber. In contrast, Indian women sat on the ground to operate the *charkha*.[29]

In the late fifteenth century, an anonymous artisan invented the flyer-bobbin wheel, also called the Saxony or flax wheel. The flyer-bobbin mechanism
further automated yarn making by imparting twist and winding on simultaneously. And it was foot-powered. The spinner sat at the wheel and treadled it,
much as the ginner treadled the foot gin. Because of the pull on the yarn exerted by the flyer-bobbin mechanism, however, a fine cotton or woolen yarn
could not be spun. Spinners used the great wheel to produce fine yarns and
the flyer-bobbin wheel primarily to make linen or worsted yarns.[30] The flyer-
bobbin wheel further mechanized yarn production but did not change the 1:1
ratio of spinner to spindle.

This ratio became an issue in the mid–eighteenth century, when the expanding fustian and cotton industries demanded yarn in quantities that spinners working one to a wheel could not meet. What the industry needed was a
machine that allowed one spinner to spin multiple yarns simultaneously.
English mechanics applied the principles of the great wheel and the flyer-
bobbin wheel to the three spinning machines they invented in the late eighteenth century. James Hargreaves based his hand-powered jenny on the principle of the great wheel but belted one wheel to eight spindles in an early
model. He patented the jenny with sixteen spindles in 1770. Richard Arkwright used the flyer-bobbin as the foundation of his water-powered spinning
frame, which was drawn in the 1769 patent with four flyer-bobbin assemblies.
Requiring varying levels of skill, both machines increased spinners' productivity but neither produced the range of yarns that textile makers needed. The
jenny spinner spun a soft yarn suitable for weft or the horizontal elements in a

fabric. The frame spinner produced a hard coarse yarn that served for the warp, the lengthwise threads, as well as coarse weft yarn.

In 1779 Samuel Crompton introduced a machine with which one spinner could produce multiple yarns in a range of sizes. His "mule" was a hybrid of the jenny and frame. He borrowed the idea of the jenny's spindle but a mounted a row of them, forty-eight on an early model, on a "wheeled carriage." The spinner pulled the carriage out to attenuate the yarn and to allow twist to run into it, and then pushed it in to wind the yarn on. Sets of differentially rotating drafting rollers, borrowed from the frame, meted out fiber in the form of a prepared roving. By 1800 steam-powered mules carried over 250 spindles. The number increased to over one thousand as carriage movements were automated, a process that culminated with the invention of the self-acting mule in 1835.[31] With the mule, spinners produced the amount and variety of yarns that velvet, satin, fustian, and cotton textile makers needed. The mule in combination with the water frame, which continued to be used to produce low-end goods, forced downstream changes, notably the expansion of cotton production and ginning innovation.

During the 1770s colonists in British North America joined their Caribbean cohorts in the cotton trade, growing fuzzy-seed cotton, ginning it with the foot gin, and shipping it to Britain. In 1768, for example, Virginians shipped 43,350 pounds, South Carolinians 3,000 pounds, and Georgians 300 pounds of cotton fiber to Great Britain.[32] Floridians contributed fluctuating amounts that ranged from a high of 40,823 pounds in 1773 to a low of 500 pounds in 1778, and then back up to 21,778 pounds in 1780.[33] Travelers confirmed the renewed interest in cotton growing. Visiting Georgia in 1774, William Bartram recorded in his *Travels* "the poorer class of people" planting cotton. They grew two types, according to Bartram, a low annual and a tall "West-Indian" perennial, and that the fiber of both was "long" and very white. He traveled up the Pearl River, between Louisiana and Mississippi, and saw cotton growing along with other commercial and vegetable crops.[34]

Bernard Romans was a contemporary of Bartram's. He not only recorded the details of the natural resources of East and West Florida but also included precise descriptions of cotton and the foot gin in his 1775 *Natural History*.[35] Romans followed the textile industry and saw in cotton the means to colonial prosperity. Colonists could "never raise too much" cotton, he wrote, since "Being so useful a commodity, that scarce any other exceeds it." He urged colonists to emulate the "industrious Acadians" and "manufacture all our necessary clothing in Florida of this staple," but not at the expense of foreign trade. "Manufactures in Lancashire, Derbyshire, and Cheshire" used it "alone or in mixture with silk, wool, flax, &c.," relying on foreign imports "chiefly from the Levant to so great an amount as near 400,000l. sterling value." Colonists could capture some of that trade, he suggested, and hoped his readers agreed. Should they find it "worthy of a more universal propagation," he added the details of "culture and of cleaning this produce of our earth," explaining how to grow and gin cotton.[36]

Romans advised colonists to cultivate "*Gossypium Anniversarium* or *Xylon Herbaceum;* also known by the name of green seeded cotton, which grows about four or five feet in height." A short paragraph on planting and harvesting preceded two on the gin and ginning. The cotton must first be harvested, he explained; then the seed cotton "must . . . be carried to the mill." The mill was a four-legged, heavily timbered, foot gin "about four feet high." It was "joined above and below by strong transverse pieces; across this are placed two round well polished iron spindles, having a small groove through their whole length, and by means of treddles are by the workman's foot put in directly opposite motions to each other."[37] Romans did not explain the function of the "small groove" along the length of the iron rollers, but it may have added friction to the smooth rollers, enabling them to better hold and pull through the fiber.

The ginner sat on a bench built into the frame in front of a "thin board of seven or eight inches wide and the length of the frame. . . . With his left hand he spreads [the seed cotton] on this board along the spindles which by their turning draw the cotton through them being wide enough to admit the cotton, but too near to permit the seed to go through, which being thus forced to leave the cotton in which it was contained, and by its rough coat entangled; falls on the ground between the workmans legs while the cotton drawn through falls on the other side into an open bag suspended for the purpose under the spindles."[38] Romans described the same type of gin that the woman used in Diderot's 1762 drawing. His exhortations and descriptions confirmed that colonial southerners were expanding cotton production as a consequence of increased British demand and that they used foot gins to process the raw material. Even as his book went to press, agitation in the North American colonies threatened to interfere with what had become an ideal mercantilist relationship.

The revolution that erupted in 1775 formally ended in 1783 with the Treaty of Paris, which conferred political autonomy to the new nation that had called itself the United States of America since 1778. The same treaty, however, also limited its economic options. The British had discontinued bounties on rice and indigo and closed Caribbean and Asian ports under their control to American ships. During the Revolutionary Era, contingency had moved cotton from the margin to the foreground of United States economic policy, as Americans substituted domestic manufactures for British imports. They had expanded cotton production and initiated industrialization. At war's end, British merchants ironically found themselves in need of a commodity that their nemesis produced at a time when American producers needed access to global markets.

Mutual need trumped political differences, as United States planters dedicated acreage to cotton for the British trade. The trade, in turn, forced a debate over development policy. If United States citizens continued to ship raw materials like cotton to Great Britain and import manufactured goods, then Britain would continue to benefit from a favorable balance of trade and the added value accrued to manufactured goods. Americans would remain economic dependants. The alternative would require a rejection of mercantilism and an embrace of capitalism. Americans would develop a domestic textile industry that would use domestically grown cotton to produce yarn and cloth for internal as well as global markets. They would centralize cotton production, processing, and manufacture in the United States, affirming their independence and challenging British hegemony. The first choice blunted the impact of independence; the second sharpened it.

The constitutional debates raged in Philadelphia as statesmen examined the role of government in the economic development of the new republic. Thomas Jefferson argued for dependence in his 1787 *Notes on the State of Virginia*. In the section on "Manufactures," he explained that Americans needed to examine the implications of the European theory that "every state should endeavor to manufacture for itself." He pointed out that necessity, not choice, had forced the "surplus of the people" into manufacturing in Europe. In contrast, tending "the immensity of land" in the United States should occupy "all our citizens." Indeed, he declared, "those who labor in the earth are the chosen people of God, if he ever had a chosen people, whose breasts he has made his peculiar deposit for substantial and genuine virtue." He continued, "Let our workshops remain in Europe. It is better to carry provisions and materials to workmen there than bring them to the provisions and materials, and with them their manners and principles."[39] The political revolution was sufficient for Jefferson. He believed that an industrial revolution—and, with it, the formation of a landless proletariat—would undermine the moral foundations of the new nation.

Federalists coupled virtue to economic independence. In 1787 Tench Coxe addressed an assembly of influential "Friends of American Manufactures," explaining the advantages of American agriculture but also the need for industrial development. Coxe would become Alexander Hamilton's assistant secretary of the Treasury and his ideas would migrate into the 1791 Congressional *Report on the Subject of Manufactures.* Importing manufactured goods, Coxe declared, was a "malignant and alarming symptom, threatening convulsions and dissolution of the political body." The immigration of skilled mechanics should be encouraged instead. They would "relieve and assist us" since "Ours will be their industry . . . and ours will be their skill." Together farmers and manufacturers would sustain the national market he envisioned and insure economic independence. A mixed economy, Coxe argued, would yield "more profit to the individual and riches to the nation" than one based on agriculture alone. "It will lead us once more into the paths of virtue by restoring frugality and industry, those potent antidotes to the vices of mankind."[40]

Often called the "father of the American cotton industry," Coxe encouraged southern planters to continue expanding cotton acreage through the 1780s.[41] Charleston merchants shipped 1,500 pounds of fiber in 1785, 32,400 in 1787, and 84,600 in 1788.[42] It was "worth double the money in America, which it sold for before the revolution," Coxe reported in 1787. Planters heeded his advice to "adopt the cultivation of an article from which the best informed manufacturers calculate the greatest profits, and on which some established factories depend."[43] United States producers supplied half a million pounds of fiber to British manufactures in 1789 and increasing amounts to the domestic industry to which Coxe alluded.[44] In the 1780s, the state of Massachusetts had incorporated three cotton textile mills in Bridgewater, Beverly, and Worcester. They were indicators that domestic demand for cotton, and gins, would increase.

During the 1780s British textile merchants imported most of their cotton from Caribbean producers but, selective in their needs, they explored other sources. Fustian makers John Hilton and William Frodsham argued before the Board of Trade and Plantations in 1786 that French cotton textile makers had an unfair advantage that the board should equalize. They complained that French Caribbeans grew Bourbon cotton, a variety of *G. barbadense,* and charged inflated prices for it. Rather than depend on the French, they wanted English Caribbean colonists to grow varieties with similar characteristics.[45] The sympathetic board acted immediately in 1786 to authorize colonial governors to offer bounties to sugar and other planters to encourage them to grow cotton but also to improve the gin.

The board took further action the following year to encourage the production of the "finer sorts of cotton in the West Indies" to meet fustian makers' needs. In 1787, it sent a Polish botanist to investigate the production methods of the coveted Surat cotton in India, and directed Levant Company merchants to buy the Persian cotton seed in Constantinople. John Hilton sought new varieties elsewhere in Asia. William Frodsham went to western Africa, where he collected seeds and fiber from areas along the Gambia River, near the Cape Coast fort, and in Accra, in present-day Ghana. "Cotton grew spontaneously" there, he reported. Some of it was "of an extreme fine quality" and had already been marketed in Bristol. Judged against the smooth-seed Brazilian *G. barbadense*, the cotton from Senegal, a variety of the fuzzy-seed *G. herbaceum*, was deemed superior.[46] The comments revealed that the merchants cared less about whether the cotton had a fuzzy or a smooth seed, whether it was more or less difficult to gin, than they did about its "fine quality."

One of their favorites was Persian cotton, another variety of *G. herbaceum*. It grew well in the Caribbean and produced a staple with characteristics they liked. Believing that planting seed from harvested crops caused species degeneration, they encouraged planters to "plant fresh seeds of the finest cottons from the other countries."[47] Thus British textile makers facilitated the transfer of Asian and African cotton varieties to the Americas, where cultivation met a specific need for fiber capable of being spun into fine yarn on mules. The transfers succeeded, as they had a century earlier, because they were made laterally within semi-tropical regions, eliminating the problem of day-length variability. The introduced varieties flowered and fruited in the Americas without an intervening period of acclimation.

As English mechanics completed the mechanization of spinning with three new machines and cotton growers expanded production and experimented with new varieties, cotton gin makers in the Americas remained conservative, producing variations on the roller principle. After the foot gin, their most notable invention was the barrel gin, which they introduced sometime in the 1770s. It was a composite gin made by belting from two to twenty-four roller gin heads to a single drive wheel, or barrel. A single ginner could turn smaller models, but animal mills or waterwheels were needed to power larger units. Romans had seen a two-head barrel gin in Florida that turned "with so much velocity as by means of a boy, who turns it, to employ two negroes at hard labour to shovel the seed from under the mill." A Mr. Krebs in Pascagoula owned another but had partly disassembled it on the suspicion, Romans believed, that Romans wanted to copy it. He was told that one of "those improving mills" turned out seventy or eighty pounds of fiber per day, or about three times the average outturn of a single foot gin.[48] Larger barrel gins in South Carolina were made up of twenty-four heads and ginned "from 6 to 800 weight of clean cotton in a day," turning out from two hundred to just over two hundred and fifty pounds of fiber.[49]

An otherwise careful observer, Romans omitted a crucial detail on ginner productivity that John Drayton, governor of South Carolina, addressed in his 1802 encyclopedia of the state. "A negro is stationed at each of these gins to feed it with cotton," Drayton wrote, "besides one who superintends the whole."[50] Drayton explained that one ginner stood at each gin head and manually supplied seed cotton to the rollers. The ginner still arbitrated outturn, which was dependent on how fast the ginner supplied the seed cotton, not on how fast the rollers turned. The barrel gin thus increased outturn through iteration, but it did not increase ginner productivity. Furthermore, it added to the danger of ginning. Metering seed cotton to the rapidly spinning rollers of a hand or foot gin was a dangerous occupation, but the ginner could stop either gin type almost immediately if a finger got caught. The oxen- and water-driven barrel gins that Drayton described could not be stopped quickly, so ginners risked serious injury. Barrel gins required a great deal of skill, labor, and capital, perpetuating a preindustrial order as they imposed an industrial pace.

Foot gins were not without their own risks and limitations. Drayton included a description of the gin in his encyclopedia, adding that with it "a negro will gin from twenty to twenty-five pounds of clean black seed cotton in a day."[51] The upper outturn limit was roughly what James Marion reported in 1747, which suggests that ginner productivity had not changed in over half a century. In 1803, Philadelphia encyclopedist James Mease published two drawings of the gin that illustrate that gin makers were trying to improve it. The engravings show a foot gin with iron rollers and flywheels, like the gin Romans documented. The flywheels in the 1803 drawing, however, were weighted with a "triple fly," each arm of which was loaded with a four-pound

Fig. 2.2. Foot Gin, 1803. A state-of-the-art machine, this gin incorporated accelerated flywheels (A), a trough for disengaged seeds (L), and a built-in seat (O). Note the single treadle attached to the U-shaped crank at the center of the machine (B). Lint cascaded down the incline to the left. James Mease, *The Domestic Encyclopedia* (Philadelphia: William Young Birch and Abraham Small, 1803), Plate 2. Courtesy, American Antiquarian Society, Worcester, Mass.

lead weight. The inventor claimed that, with the weighted flywheel, a ginner could turn out "sixty-five pounds of cotton per day" (fig. 2.2).[52]

Like the Chinese hand-and-foot gin, the weighted foot gin required ginners with strength and stamina. They were "always the primest young Hands upon the place," a Bahamian planter reported in 1794.[53] But even they were prone to injury. Another planter reported that the foot gin was "apt to rupture ne-groes and always injures them, if continued for any length of time."[54] Thomas Spalding, a Georgia planter, politician, and inventor, accepted as received wis-

dom that "the continued motion of the feet"—in other words treadling—
"produced a relaxed system" in pregnant women which induced "abortion or
miscarriage." He therefore used only male ginners in the 1830s and reported
no injuries.[55] Without citing specific incidents, the planters described the consequences of severe abdominal strain—even hernia—which must have been
excruciatingly painful and incapacitating. Without other evidence, however,
it is difficult to understand how or with what frequency the injuries occurred.
They could have been caused by overweighted flywheels or by unconscionable planters pushing ginners beyond their physical limits. However, if the gin
routinely caused debilitating injuries, profit-conscious planters would have
sought alternatives. Instead of being replaced, the foot gin outlasted the barrel
gin and remained in use to the mid-1860s. Nevertheless, the reports of injuries
cannot be ignored. They signify a callous disregard for slaves' well-being, as
planters admitted witnessing the painful accidents.

In 1788 Joseph Eve invented an inanimately powered, self-feeding roller
gin that promised to eliminate the risk of injury and to increase ginner productivity. He had responded to the bounty that fustian makers Hilton and
Frodsham requested and that the Board of Trade and Plantations approved in
1786.[56] This gin was another iteration of the pinch principle but moved the
evolution of the gin forward because it automated supply. Unlike the barrel
gin that maintained the 1:1 ginner/gin ratio, Eve's gin uncapped outturn by
eliminating the ginner. An attendant needed only to put seed cotton on the
tines of the toothed feeder and collect the fiber. The large gin was labor-saving
but capital-intensive. It was more expensive than the foot gin and required a
water- or windmill as well as a reinforced building that could withstand the vibrations of the feeder. Yet it increased outturn without changing the quality
of the fiber, a characteristic of the gin that endeared it to planters and textile
makers alike (fig. 2.3).

The inventor identified himself as "Joseph Eve of Pennsylvania" on his
1803 patent drawing but he spent his formative years in the Bahamas. The
first gin maker in the Americas for whom a body of evidence exists, Eve was
also one the most gifted. He was born in Philadelphia on May 24, 1769, to
Anne (Moore) and Oswell Eve, the youngest of four surviving sons and one
daughter, Sarah, who was engaged to Benjamin Rush but died tragically before the wedding. As a schooner captain, the patriarch sailed between Philadelphia and Lisbon, Jamaica, and St. Kitts in the 1750s. He was appointed war-
den of the Port of Philadelphia in 1766, the same year that he joined the

COTTON GIN,

Invented by Joseph Eve
of Pennsylvania.

Fig. 2.3. Self-Feeding Roller Gin, Joseph Eve, Bahamas, 1788. Eve left Philadelphia in 1774 but always considered himself a Pennsylvanian. The drive wheel shown at the right propelled the ginning rollers as well as the self-feeder, which was governed by the arm extending diagonally from the wheel. The illustration was taken from Eve's 1803 patent, reprinted in *The Portfolio* n.s. 5 (March 1811): 185. Courtesy, American Antiquarian Society, Worcester, Mass.

Society for the Relief of the Poor, Aged and Infirm. Marriage and a growing family grounded him after 1769. His correspondence with Benjamin Franklin and with Richard Bache identifies Oswell Eve among the elites of the city, but he made his name as a black powder maker. In 1774, a year before King

George III forbade shipments of black powder to the thirteen colonies, he built a large, water-powered manufactory on Frankford Creek, just north of Philadelphia.[57]

The mill was a technological marvel that attracted tourists like Charles Willson Peale and patriots like Paul Revere, both of whom visited, and planter-politicians like John Dickinson, who wrote Eve. On January 11, 1776, Oswell Eve and George Losch signed an "Agreement" with the Committee of Secrecy of the Continental Congress to "manufacture all the Salt Petre which shall be delivered to them severally for one year from this date into good Gun Powder." From the petition that Oswell Eve filed with the Committee of Safety for the Province of Pennsylvania on March 22, 1776, it appears that the partners had delivered the powder but that Congress had failed to compensate them. There is no evidence that disagreements over payment prompted what followed, but when the British occupied Philadelphia and took over the Frankford mill in September 1777, it was rumored that Eve cooperated. The Supreme Executive Council branded him a traitor but allowed him to leave the city with his family when the British evacuated in the spring of 1778. He left his son Oswell Jr. in charge of the manufactory.[58] Joseph was nine years old.

The son of a sea captain, inventor, and mechanic who corresponded with American *philosophes*, Joseph grew into a gifted man who was equally comfortable with the lathe, the plow, and the pen. He was nineteen in 1788 when he announced the invention of a "Cotton Machine" in the *Bahama Gazette*. The gin was the product of "considerable Labour, Time, and Expence," he informed "the Planters and Merchants of the Bahamas." It was also "an effectual, useful Machine" but inexpensive "in Comparison to its Utility." With "the Assistance of a Horse, or some other foreign Power, as the Wind, etc. . . . it will clean three hundred Pounds [of seed cotton] to a Hand per day (or one hundred pounds of fiber)," four times the average foot gin outturn. He planned to equip it with metal rollers to eliminate the need for a carpenter to turn replacement rollers, as they did for foot gins. "It shall be as easily kept in Order, and with less Expence than the common Ginn," Eve promised. But conditional and future tenses dulled the impact of the notice of the new gin. They suggested that while Eve may have built a working model, he had not put the gin into production. Indeed, the notice appeared to be a solicitation for funds to develop the idea further. Eve ended it with a request for the interested "Public" to present "a Bond or Bonds for one thousand Pounds Sterling" in order to inspect what might have been a working model.[59]

Eve was either distracted or discouraged because he fell silent for the next two years. It was John Wells, the Loyalist editor of the *Bahama Gazette,* who nudged him back into gin making and planters into patronizing him. In a December 24, 1790, editorial, Wells reminded readers that "A gin for cleaning Cotton, invented and constructed by the ingenious Mr. Joseph Eve, of Cat-Island" was on display in Nassau. It came "highly commended" by those who had used it and could be "worked with the same ease by one person, as the Gin commonly used." It had the "singular advantage of the rollers not being liable to be broken" because they were made of metal. Wells cited comparative outturn figures to whet planters' appetites. A foot ginner turned out between twenty-five and forty pounds of fiber, he estimated. From trials with Eve's gin, "one person can, during the same time, clean from eighty to an hundred pounds of Cotton." He urged planters to adopt it, concluding, "An improvement in Mechanics, of a nature so important to the Cotton Planters will unquestionably meet with its due share of their patronage."[60]

Eve addressed planters directly on the first day of 1791, from the first column of the first page. He refused to imagine himself forgotten, he began, and so wrote on the assumption that readers remembered his 1788 announcement. Planters had been less than enthusiastic, he supposed, because the gin required an external power source. They must also have thought it too complicated, because Eve added that he had tried to "simplify its construction." That proved difficult, he admitted, because "I had to improve a Machine, already highly improved, without adding much to its complication or price." Believing that he had perfected the gin, Eve put it before the planters confident that it would earn their approbation. It had already earned the approval of fourteen "subscribers and planters" who signed two appended testimonials. Eve trusted that they and other buyers would honor his rights even as he tried to secure patent protection. No conditional tenses shadowed the 1791 notice. Eve sold the gins for "32 dollars each."[61]

The testimonials that celebrated the simplicity of Eve's gin misled gin buyers and, later, scholars. Eve had identified three problems with the foot gin: its size, manual supply, and source of power. Ambitiously addressing all three, he built a frame in the form of a right triangle about four feet long and five feet high. He added fluted "rollers of wood or metal, which draw the cotton through perpendicular apertures too small to admit the seeds." Somehow compensating for springing, he made the rollers approximately five-eighths of an inch in diameter and thirty inches long, locating them on the hypotenuse of the frame. A complicated self-feeder sat over the rollers. According to the 1803 patent description, "There are three sets of metal combs placed before the rollers . . . that feed the machine and disengage any extraneous matter; the middle set move up and down and pass between the teeth of the others."[62] In 1828 Basil Hall described it as "a sort of comb fitted with iron teeth, each of which is a couple of inches in length and seven-tenths of an inch distant from its neighbour. . . . This rugged comb . . . lies parallel [to the rollers], with the sharp ends of its teeth almost in contact with them." It was made "to wag up and down with considerable velocity, in front of the rollers . . . opening the seed cotton before the rollers pulled in the fibers.[63] An eyewitness, Hall watched the seeds fly off "like sparks, to the right and left, while the cotton itself passes through between the rollers."[64] The patent description explained the doffing process. Brushes "fixed behind the rollers through the whole length of the feeding spaces," removed the fiber, preventing it from "wrapping round them or jamming." By any measure, this was a complicated machine.[65]

Bahamian planters mastered the gin's complexities and in 1794 published unsolicited testimonials. Intentionally biased against the foot gin, the testimonials offer valuable insights on gins and ginning during the era of the roller gin. "A Cotton Planter," the same who described "rupture[d] negroes," submitted one of the more detailed accounts, based on a harvest of ten tons of seed cotton, one half of which was processed with one of "Eve's Wind [powered] Gins" and the other with foot gins. He paid £132.4 for Eve's wind gin, which he calculated would turn out "250 lbs. per day," ginning five tons in "40 days," hired "2 hands" at "4s each," and spent £1.10 for ten pairs of wooden replacement rollers, apparently deciding against iron rollers. Presumably, he already had a windmill and a building for the gin, since he did not mention either expense. He pitted the "wind gin" against twelve of the "common foot gin" purchased at a cost of "18 dollars each," or £86.8. A carpenter kept "the gins in order, and making rollers, at 6s" per day for thirty-three days. Wages for twelve ginners at 4s. per day for 400 days, the length of time he estimated it would take them to gin five tons of seed cotton at an average daily outturn of "25 lbs per day," came to £80. He concluded that "eighteen hundred days are saved by" using Eve's wind gin, at a time in the planting year when "the slave's work is most valuable. This labour in good years will raise two tons of cotton."[66]

"A Cotton Planter" used the numbers to impress on readers the rationality of Eve's gin and the irrationality of the "common Foot Gin," but he also complicated the picture of the single ginner working alone that Diderot had pic-

tured and Romans had described. Bahamian planters used teams of foot ginners, a practice that would be documented on the mainland in the next
century. The use of teams of foot ginners undoubtedly inspired the invention
of the barrel gin, which belted multiple gin heads to one power source. His account also highlighted the contingent costs that gins incurred. The foot gin
required the labor of carpenters to turn replacement rollers, and the Eve gin
required attention from mechanics, a point that mainland planters would
later emphasize. Yet they would also echo the author's conclusion. Foot ginners could be productive, turning out upwards of "40 lbs, some say more," but
they were the most valuable men that a planter owned or hired. With Eve's
gin, planters used that labor to produce cotton and "only the most ordinary"
workers to gin it.[67]

News of Eve's gin reached the mainland in February 1791, when the Savan-
nah-based *Georgia Gazette* reprinted John Well's 1790 editorial.[68] In the spring
of 1796, Thomas Spalding published a notice soliciting orders for the gin. Like
Eve, Spalding was the son of a Loyalist but fourteen years his junior. Eve "empowered" him to receive subscriptions for the gins and offered to deliver them
to Savannah for "50 guineas," without the power source. Mainland planters
cultivated both smooth- and fuzzy-seed cotton varieties, and Eve assured
them that the gin was effective on both staples. He promised an average outturn of three hundred pounds of smooth-seed cotton and from one hundred
and fifty to two hundred pounds of fuzzy-seed varieties. Spalding verified the
amounts and added that the gin had met "universal approbation in every part
of the West Indies to which it has been sent."[69]

Ginning technology was not a bottleneck as the nineteenth century
dawned. Producers in the Caribbean and on the mainland used human-
powered foot gins and water-, wind-, and animal-powered barrel and Eve gins
to process their cotton crops. Southern planters shipped nearly half a million
pounds of fiber to the British cotton industry in 1790 but also supplied the
fledgling American industry, which had developed despite congressional rejection of Hamilton's *Report on Manufactures*.[70] Three factories opened in
1790 alone. British mechanics William Pearce and James Hall managed different departments of Hamilton's Society for Establishing Useful Manufactures, a
cotton and wool textile manufactory in Paterson, New Jersey. Samuel Slater
had arrived from Derbyshire and had put merchants William Almy and Smith
Brown's spinning frame in working order in Pawtucket, Rhode Island. Hugh
Templeton operated a water-powered textile mill in upcountry Stateburg,

Sumter District, South Carolina, that combined ginning, likely with an Eve or barrel gin, carding, slubbing, and spinning. He used hand-powered jennies that totaled eighty-four spindles and in 1789 unsuccessfully petitioned the state legislature to expand the scope of the firm by building a patented carding and cotton spinning machine.[71] The rising cotton export figures and the growth of the domestic cotton industry validated American gin makers. They had modernized the medieval roller gin yet remained faithful to the pinch principle, building gins that delivered cotton fiber with the qualities textile makers valued. They could not have anticipated the challenge launched in 1794 by a new type of gin that violated the principle and destroyed the qualities.

CHAPTER THREE

The Invention of the Saw Gin, 1790–1810

In 1794 Eli Whitney patented a new ginning principle and built a new kind of gin. Instead of rollers that pinched off the fiber, he used coarse wire teeth that rotated through a tightly spaced metal grate to pull it from the seed. Like Eve's self-feeding roller gin, it required only an attendant to supply the seed cotton; unlike Eve's gin, it privileged quantity over quality. It turned out large quantities of fiber but destroyed the very qualities that textile makers had valued. Whitney's gin shattered apparent equilibrium in the industry. It forced textile makers to reevaluate their concept of quality, and producers their commitment to the roller gin, as it instigated a regionally shaped response by gin makers. Antipathy toward the wire-toothed gin invigorated roller gin invention in southeastern states. Industry ambivalence spurred others to adopt the gin but change it. Gin makers substituted an axle loaded with fine-toothed circular saws for Whitney's wire-studded wooden cylinder. In 1796 Hodgen Holmes of Augusta, Georgia, patented the adaptation, naming it the saw gin. The suit that followed capped a contentious and socially and legally mediated process from which Eli Whitney emerged as the inventor of the cotton gin.

Judges construed the saw gin as a turning point in southern history, even as the textile industry remained committed to the roller gin.

The introduction of a new gin in 1794 was as unexpected as it was unprecedented. It was unexpected because the British textile industry had expanded from the sixteenth through the eighteenth centuries without a change in the ginning principle. Cotton producers had increased the acres they planted in cotton and planted new varieties to suit textile makers. The market attracted new producers who, like established planters, used roller gins to process their crops. Roller gins, whether hand-cranked in the Levant and India, or foot-, animal-, and inanimately powered in the Americas, provided adequate amounts of fiber with the qualities that textile makers wanted, namely length and cleanliness. All roller gins removed the fiber by pinching it off in bundles, preserving its length and orientation as grown. Random fragments of fractured seeds were picked out of the fiber before it was bagged and shipped.

In 1788 Joseph Eve gave planters and merchants a machine that bridged the medieval and modern. It preserved the ancient roller principle but completed the appropriation of the ginner's skill, as Arkwright's frame had that of the spinner. Appropriation had proceeded in stages beginning with the single-roller gin that mechanized the thumb and finger pinching motion. The roller gin in turn appropriated the agility and strength needed to manipulate the single roller, while the foot gin freed both hands to supply seed cotton. The barrel gin used animal and water power, removing humans as a power source but retaining them as seed cotton suppliers. The self-feeding animal-, wind-, or water-powered Eve gin replaced each of the skilled tasks of the ginner with mechanical components.

Nevertheless, Eli Whitney's unprecedented gin filled a vacuum. While large merchants invested in barrel gins and large planters in the Eve gin, the majority continued to use the skill- and labor-intensive foot gin to gin fuzzy-seed, short-staple cotton as well as the smooth-seed, Sea Island cotton. Barrel gins had not decreased the number of ginners and only marginally improved ginner productivity, and Eve's complicated gin was notoriously finicky. Whitney ignored these modernizing gins and offered a replacement for the ubiquitous foot gin. As he wrote in his 1793 petition for a patent, his gin was "entirely new & constructed in a different manner and upon different principles from any other Cotton Ginn or Machine heretofore known or made for that purpose." Furthermore, "with this Ginn, if turned with horses or by water, two persons will clean as much cotton in one Day, as a Hundred persons could

clean in the same time with the ginns now in common use." Whitney concluded the petition with the claim that "the Cotton which is cleaned in this
Ginn contains fewer broken seeds and impurities, and is said to be more valuable than Cotton which is cleaned in the usual way."[1] This was not another iteration of the roller principle, Whitney made clear, but an "entirely new" ginning principle that promised to uncap outturn. Two ginners on the new gin
could turn out as much fiber in a day as one hundred foot ginners each averaging twenty-five pounds, in all twenty-five hundred pounds, Whitney estimated. Specialists had the fiber examined and deemed it cleaner and more
valuable than that from roller gins.

Just as the petition belied Eli Whitney's later statement that the idea of the
gin sprang from his mind "involuntarily," the patent specification revealed an
individual who was a deliberative inventor.[2] Fulfilling the terms of the 1793
Patent Act, Whitney submitted both a long and a short description, today
called specifications, to the Office of the Secretary of State. He addressed the
cover letter to Thomas Jefferson—examiner of all patents, as secretary of
state—and mailed the three documents on October 15, 1793, apologizing for
the delay of nearly four months since filing the petition. He expressed the
hope in the letter that what the long description lacked in eloquence, it com-
pensated for in accuracy.[3] Both survived as official copies signed and sealed by
President James Madison in November 1803. John S. Skinner, the editor of the
American Farmer, reprinted the short description in 1823.[4] He also published
the only extant original drawing of the 1794 patent (fig. 3.1). The drawing and
the descriptions revealed the revolutionary nature of the new fiber removal
technology.

Conceptually and structurally, Whitney divided the gin into five "principle
parts," the frame, the cylinder, the breastwork, the clearer, and the hopper.
The frame alone differentiated it from the roller gin. Like flatbed carding engines of the 1790s, Whitney's gin presented itself as an impenetrable rectangular box that contained the working parts. The four "parts" were similar to
those in other textile machines, but Whitney had reconfigured and rear-
ranged them. The cylinder and clearer were cognates of the rollers of the roller
gin. Like the rollers, they rotated in opposite directions and in close proxim-
ity. But Whitney enlarged them and changed their functions. The cylinder,
labeled "B" in the patent drawing shown in figure 3.1, was the only one of the
two actively engaged in ginning. Made of solid wood, it was six to nine inches
in diameter and two to five feet in length. Whitney studded it with coarse

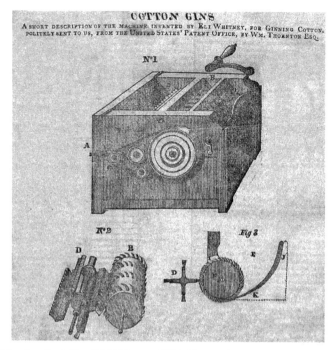

Fig. 3.1. Wire-Toothed Gin, Eli Whitney, 1794. This is the only extant patent drawing of Whitney's original gin. When a fire destroyed the patent office in 1836, Congress approved funds for patentees to restore lost files. The drawings that Whitney's heirs restored (Whitney died in 1825) differed considerably and included saw teeth as an alternative construction technique. *American Farmer* 4 (21 February 1823), 380.

iron teeth arranged in annular rows spaced by the size of a cotton seed.[5] The clearer, labeled "D," did not assist in the removal of the fiber from the seed. It brushed the fiber from the teeth of the toothed cylinder and whisked it away from the frame. The breastwork and hopper were the heart of the new ginning principle. They acted with the toothed cylinder as the second half of the ginning mechanism. The breastwork was a grate, which Whitney made out of metal into which he cut slots wide enough to accommodate the teeth but too narrow for a cotton seed to pass through. He attached it to the back of the hopper, the receptacle for the seed cotton.

A description of the gin in motion illustrates how Whitney harmonized the five principle parts. He belted the wire-toothed cylinder to the power source and thence to the clearer, arranging "whirls" (pulleys) such that the cylinder and clearer turned in opposite directions and that the clearer rotated "much faster than that of the cylinder." When power was applied, the wire teeth revolved through the seed cotton in the hopper, starting the ginning process. The teeth pulled both fiber and seed along until they reached the slotted breastwork at the back of the hopper. At the moment the teeth passed through the breastwork and pulled the fiber through it, ginning occurred. The breastwork held back the seeds and allowed the revolving teeth to remove the fiber from them. The rapidly revolving clearer brushed the fiber from the teeth, and, in Whitney's words, "the centrifugal force of the cotton disengages it from the brushes" and away from the gin.

The gin was a technical marvel that Whitney described with precision in the long patent description. He made reference to other textile machinery, mechanical principles, and standard shop practice, particularly regarding the construction of the cylinder and clearer. Whitney built both around an iron axle and specified how they were to ride in the "bearing or box[es]" that held the lubricant. The paragraph on the manufacture of the wire teeth included advice on wire selection and directions for flattening, cutting, and inserting the wire into the wooden cylinder, which he accomplished with a "light hammer." The teeth were then cut even "with a pair of cutting pliers" and the cylinder secured in "a lathe" so that the gin maker could file the teeth "to a kind of angular point" and finish them with "a polishing file."

Directions on making the breastwork were equally detailed. Whitney understood that the angle at which the teeth passed through the breastwork was crucial, and that passage would "give the [seed] cotton a rotary motion in the hopper," forming what ginners later called the seed roll (fig. 3.2). He calculated the thickness of the breastwork from the average length of cotton and the spacing of the grate and teeth from the average size of a cotton seed. The clearer received similar fastidious attention. Whitney likened its construction to that of a "weaver's sleigh" or reed, a device that spaced the warp yarns and that the weaver used to beat the weft into place. After describing various ways to build the arms and attach the bristles, Whitney emphasized that the distance between the arms was critical. It "must be at least 4 or 5 inches; otherwise the cotton will wind up around the clearer." The angle at which the clearer hit the cylinder was also specified. Whitney wrote that it should be placed so that the "bristles will make an angle of 20 or 25 [degrees], with the diameter of the clearer . . . by that means the bristles fall more perpendicularly on the teeth, strike them more forcibly, and clear off the cotton more effectively."[6]

Fig. 3.2. The Seed Roll. Whitney accurately described the phenomena of seed roll formation, pictured here in a modern saw gin. Note the agglomeration of seeds at its core. Courtesy, USDA, ARS, Southwestern Cotton Ginning Research Laboratory, Mesilla Park, New Mexico.

That Whitney had studied cotton and the gin was evidenced by his final comments. "Black seed cotton" could be ginned continuously because the seeds stripped clean and full through the hopper. But the "green seeds, which are thus denominated from being covered with a kind of green coat, resembling velvet will continue in the hopper." The seed roll would grow in size until the seeds interfered with the teeth of the cylinder. As a solution, he recommended that one hopper of seed cotton be ginned at a time, with "the movable part drawn back, the hopper cleared of seeds and then supplied with cotton anew." The gin ran intermittently but quickly. "It will diminish the usual labor of cleaning the green seed cotton by at least forty nine fiftieths," he estimated.[7]

Unlike Joseph Eve, the son of a Loyalist who had undoubtedly worked around his father's mills, Eli Whitney appeared to have few of the skills that would have equipped him to invent so complex a machine. The son of a pa-

triot, Whitney was born in 1765 to prosperous descendants of English immi-
grants who had settled in Westborough in central Massachusetts. How he
spent his early years has been obscured by the Weems-like anecdotes published by Denison Olmsted in his 1832 biography of Whitney. These describe
Whitney as a precocious boy who constructed a violin, reassembled his father's watch, repaired his step-mother's valuable knife, and designed machinery to make nails, machinery that he later reused to make hatpins and walking
canes.[8] The stories may contain elements of truth, but it is noteworthy that
Whitney never invoked them. They were constructed after his death in 1825
and were designed to reshape his past to predict his noteworthy future. Such
celebratory narratives are common. They do not intend to mislead but to convey respect and admiration. They are, however, unreliable as evidence. The
evidentiary trail begins in 1785, when Whitney was twenty years old, revealing a young man immersed in the world of the mind.

From 1785 until he matriculated at Yale College in 1789, Eli Whitney alternated winters teaching school in and around Westborough with summers
studying at Leicester Academy in nearby Leicester, Massachusetts. Yale College alumnus Ebenezer Crafts established the academy in 1782 with Jacob Davis "for the purpose of promoting piety and learning" so that the "farmer's
son might have the education of the merchant's son."[9] Crafts influenced
Whitney's choice of Yale over closer Harvard College by arranging private pre-
paratory studies for him with the Reverend Elizur Goodrich in Connecticut.
Once considered for the presidency of Yale College, Goodrich was an Episco-
palian minister and noted mathematician who polished Whitney's intellect
and manners. Whitney's exemplary references from his teaching positions,
his Leicester Academy affiliation, and his association with the Goodrich family assured him an audience with Yale president Ezra Stiles. His mastery of
Greek, Latin, and English grammar, scripture, and mathematics impressed
Stiles and secured Whitney a place at Yale in 1789.[10]

Whitney flourished during his three years at Yale, earning accolades from
President Stiles that would elevate him among his classmates. Stiles, whose
high standards were legendary, asked Whitney to deliver a 1792 graduation
oration, which was published in the same year as *An Oration on the Death of
Mr. Robert Grant, a Member of the Senior Class, in Yale-College, Connecticut.*[11]
Stiles further rewarded him with a plum tutoring position in the South. He
had done the same for Connecticut-born Phineas Miller upon his graduation
in 1785, sending him to Georgia. Since then, Miller had tutored the children

of Revolutionary War General Nathanael Greene and his wife, Catharine, who lived on the confiscated Loyalist plantation Mulberry Grove, near the Savannah River. Stiles asked his former graduate to introduce his newest to southern society.[12] Both eldest sons in their late 20s forging futures far from home, Miller and Whitney began what would become a long and complicated relationship.

The earliest extant correspondence between the men set a precedent that would persist. Chronically short of funds, Whitney asked Miller for a loan to cover the price of the packet ticket south, and Miller answered by stipulating the conditions of the loan.[13] Whitney wrote from the Goodrich residence in New Haven on the 18th of October, 1792, to Miller in New York, who answered two days later. Thanks to the loan, Whitney was able to join him and the Greene family in New York, sailing with them to Savannah. It was a tortuous adventure for Whitney. He wrote his father on the first of November, 1792, that he had been "sea-sick for six Days" but had finally arrived at "the late General Greene's Plantation, which is on the river Savannah twelve miles above the town."[14] Catharine Greene had been widowed in 1786 when General Greene died suddenly of what was diagnosed as sunstroke. Miller now managed the Greene estate, which was primarily a rice plantation, but he may have intended to diversify into cotton. Edward Rutledge assumed as much when he wrote him in 1790 for advice on cotton seed oil production and marketing.[15]

Eli Whitney spent the last days of October 1792 at the Mulberry Grove plantation, prepared to leave on November 2 to take the job that Dr. Stiles had arranged. It would inaugurate the life for which Whitney had trained. On the day before, Whitney wrote his father reassuring him that he considered it his "duty" to keep him informed him of his plans and that he would communicate "as often as possible." He planned "to go into Carolina, thirty miles from here, and enter on my tutorship" the next day and promised to forward the new address "as soon as possible."[16] As was his habit, on the same day Whitney addressed a similar letter to his friend and colleague Josiah Stebbins. Correspondence between the men then ceased. It resumed five months later on April 11, 1793, when Whitney wrote both his father and Stebbins, explaining that for various reasons he had not taken the tutoring position. He apologized to his father for the problems his debt to Mr. Goodrich had caused and complained to his friend, "Fortune has stood with her back towards me ever since I have been here and when she turns round whether she will frown or smile is quite uncertain." Homesick and disappointed, Whitney promised both men that he would be back in New Haven in June at the latest, "as poor as I came."[17]

Twenty days later, on May 1, 1793, Eli Whitney wrote Stebbins a letter that initiated the mystery and the myth surrounding the invention of the wire-toothed gin. Intimating more than he revealed, Whitney wrote that "Dame Fortune" had finally turned and smiled. He "had become very expert in the Hocus Pocus line," he jibed, and had "only two grades more to rise before [he would be] Chief of all the Magicians." Drawing a salary that he expected would soon double, Whitney planned to "pay all my honest debts and do as I please with the rest."[18] Phineas Miller explained Whitney's obscure references in a letter to Thomas Jefferson on May 27, 1793, in which he introduced Whitney as the inventor of "a machine for ginning cotton."[19] Three days later, Miller reached Whitney in New Haven, where he was making arrangements to patent and manufacture the "machine." Miller urged him to work quickly because there were "two other claimants for the honor of the invention of cotton gins, in addition to those we knew before."[20]

Consciously and collaboratively, Eli Whitney and Phineas Miller obscured the origins of the gin and laid the foundation for the myth that developed about it. The correspondence portrays a dispirited Whitney, prepared to return home on April 11 and suddenly reversing his fortunes twenty days later. Three weeks later Miller wrote Jefferson, secretary of state and a personal friend of the Greenes, explaining that Whitney had, over the previous winter without the benefit of "tools or workmen," invented a gin that "deserves the encouragement of the Public." Three days later Miller urged Whitney to make haste, reminding him that he was in a race against two unknown and several known inventors. Deliberative invention on Whitney's part seems to have played no part in the invention of the gin. Instead the men seemed engaged in duplicity masked by prestige, Miller reminding Jefferson that Whitney's "amiable character has a particular claim to private friendship & patronage." Over time, the men would deepen the mystery of the gin's origins as they re-shaped Whitney's personal history.

Whitney began on September 11, 1793, with a letter to his father that ig-nored the despondent April 11, 1793, letter and offered a new explanation for the events of the winter of 1792. At Mulberry Grove, he had overheard "very respectable Gentlemen" discussing the need for a machine that could "clean the cotton with expedition," Whitney began. Some time later, he "involuntarily happened to be thinking on the subject and struck out a plan of a Machine in my mind." He shared the plan with Phineas Miller, who promised to finance it completely should it prove successful. In the meantime, Whitney explained, he learned that the tutoring position paid half of what he had expected. Knowing his school loans had given his father "some trouble," Whitney added that he had waited until he was sure the idea would be profitable before refusing the job. Then, "In about ten Days," he had built a model for which he was offered twice what the tutoring job would pay. He spent all his time "perfecting the Machine" and succeeded in building a model "that required the labor of one man to turn it." Swearing his father to secrecy, Whitney then shared his plan to patent the gin in the United States and England.[21]

Problems with constructing the patent model delayed Whitney's plans and deepened the mystery of the gin's origins, as they kept Whitney on the shop floor systematically experimenting with materials and techniques. As late as December 1793, six months after he had mailed Jefferson the petition and nearly two months after he had submitted the descriptions, Whitney had not shipped the required working patent model. He confided to Stebbins that the toothed cylinder was the cause of the delay. Whitney had made it out of a solid piece of wood with a center bore for the iron axle. When he hammered the metal teeth into the specified dense configuration, the wood split. Whitney first reasoned that the splitting problem lay in the type of wood he used but, rethinking the problem, considered the wire-insertion method again. He had previously driven the wires parallel to the grain. Now he drove them "across the grain so that it cuts its way and does not strain the timber."[22] Whitney incorporated the results of his apparently successful experiment in the "Remark" paragraph in the long description, sharing his conceptual progress along with the required manufacturing alternative. There may have been other manufacturing problems that Whitney faced, because he did not complete the model until the spring of 1794. The patent was awarded on March 14, 1794, retroactive to November 6, 1793, when Jefferson received the descriptions and drawing.

Phineas Miller's business acumen matched Eli Whitney's persistence. With the resources of Catharine Greene's estate, Miller funded the development of the gin, including the construction, staffing, and rental or purchase of Whitney's manufactory in New Haven.[23] In return, Whitney signed a "Deed of Transfer" on the day he received the patent, assigning to the Greene estate half the value of the "right, title, claims, and interest" through Miller.[24] The men formalized their partnership on June 21, 1794, thereafter doing business as Miller & Whitney.[25] Rather than sell or lease the gins, or sell licenses authorizing others to make, sell, and use the gins, Miller initiated a plan that would allow the partnership to retain all rights to the invention. He would install the gins in ginneries under his control and charge a fee, or toll, for the service of ginning. At the larger ginneries, he planned to install cotton seed oil presses.[26] The presses would cap Miller's plan to create an integrated cotton-processing business, one that combined gin manufacturing with the services of ginning and cotton oil production. He directed all of his energies toward this remarkably ambitious vision.

Anticipating the patent award, Miller placed a notice announcing a new "Cotton Ginning" service in two Georgia newspapers in the first week of March 1794; this advertisement would add another element to the myth of the gin. The notice in the *Georgia Gazette,* published out of Savannah, reached coastal planters; the *Augusta Chronicle* circulated upriver. Rather than describe the new gin, Miller characterized it by the quality of fiber it purportedly turned out. "The subscriber will engage to gin," it began, "in a manner equal to picking by hand, any quantity of green seed cotton." Miller compared Whitney's gin to finger ginning, implying that the gin delivered undamaged, clean fiber, free of the seed fragments that had to be picked out of roller-ginned cotton. He did not intend a literal translation, that finger ginning was the only alternative to Whitney's gin. The cost was high, "viz, for every five pounds delivered him in the seed he will return one pound of clean cotton fitted for market." Five pounds of seed cotton yielded approximately one and one half pounds of fiber; Miller would return one pound and keep the half-pound. Finally, he promised to have "ginning machines to clean the green seed cotton . . . erected in different parts of the country before the harvest of the ensuing crop." As advertised, ginneries opened in Georgia and South Carolina in the fall of 1794, diffusing the new principle and fiber with new characteristics.[27]

Miller's notice cleverly avoided allusion to the roller gin makers with whose newspaper notices his competed. Most were carpenters and millwrights who made foot and barrel gins on an expedient basis. George Powell, a furniture maker, sent a notice to the *Augusta Chronicle* on March 25, 1790, announcing the continuation of his "Chair Making Business in its various branches" at a new location. He added a postscript soliciting "Gentlemen in want of Cotton Gins," informing them that they were available and built "upon a new construction."[28] Joseph Eve was known in Miller's target area. In

February of 1791, the editor of the *Georgia Gazette* reprinted John Well's editorial announcing the availability of the self-feeding and -powered gin.

Northern mechanics also made gins and advertised them in southern newspapers, appealing to planters and alerting southern mechanics. William Pearce, a British mechanic at Alexander Hamilton's Society for Establishing Useful Manufactures in Paterson, and his associate Thomas Marshall, attracted Thomas Jefferson's attention in the early 1790s. Notices of their twelve-head barrel gin appeared in newspapers in New Jersey, Philadelphia, New York, and Georgia in December 1792, claiming that it turned out 300 pounds of fiber in a twelve-hour day.[29] After reading one of them, Jefferson wrote Pearce. He indicated that ginning was "one of our greatest difficulties in the course of our houshold manufacture in Virginia," assuring him there would be "much interest in this discovery."[30] He invited Pearce to visit him in Philadelphia but had not received a reply before Whitney's patent drawing arrived on the sixth of November. In his acknowledgment, Jefferson queried Whitney: "Is this the machine advertised the last year by Pearce at the Paterson manufactory?"[31] Whitney admitted that he had read a notice in a "Savannah newspaper" sometime "last spring" but denied that his gin bore any resemblance to the "machine advertised by Pearce," which was merely "a multiplication of the small rollers used in the common gins." He had not invented another barrel gin, Whitney assured Jefferson, adding that he had not seen it or "a machine of any kind whatever, untill after" he had invented his.[32]

George Powell, William Pearce, and Thomas Marshall may have numbered among the "claimants for the honor of the invention of cotton gins" alluded to by Phineas Miller in his May 31, 1793, letter to Whitney, but there were also anonymous mechanics who supplied various types of gins to planters and merchants. Enslaved African mechanics undoubtedly made the foot gins that were used on most plantations and may have worked in shops where barrel and Eve gins were built. Southern white mechanics who made gins on an ad hoc basis, never advertising the service, also remain anonymous. Even those who published gin notices rarely included descriptions that identified the types of gins they made. Like Miller, advertisers of gins or ginning services focused on fiber qualities, quantities, and costs. Scholars are forced to extrapolate the number of gin makers and to infer from the language and context of newspaper notices the types of gins that were made.

Age was the critical identifier of gin type. Two notices serve as examples. In the spring of 1793 David Robinson, a merchant ginner whom Phineas Miller would later hire to manage one of his ginneries, placed a notice in the *Augusta Chronicle* soliciting "a few young Negroes for ginning cotton."[33] William Kennedy ran a similar notice in the *Chronicle* through the late winter and early spring of 1794. He needed to hire "About twenty Negroe boys from 8 to 10 years of age, to attend a Cotton Machine."[34] Foot gins required the labor of adults while children could tend "cotton machines," which referred to an animal- or inanimately powered barrel or Eve gin, and later to the wire-toothed gin. Robinson's notice was admittedly ambiguous but that both merchants sought young enslaved Africans suggests that they ran "machines." Children could keep these gins supplied with seed cotton and sweep away the seed, but they were not strong enough to operate foot gins. Kennedy made the distinction apparent in a 1797 notice when he advertised for "a few Negroes to work foot Gins, also some boys or girls to attend a cotton machine."[35] He did not specify what "machine" he used or why he used it as well as foot gins, only that he needed two distinct groups of workers. The notice evinced another reality—that merchants, ginners, and gin makers in up-country Georgia remained committed to the roller gin after Miller introduced the wire-toothed gin.

The momentum of roller gin culture did not daunt Phineas Miller, who proceeded on schedule to implement his plan. He received the first shipment of wire-toothed gins at the Mulberry Grove plantation on May 11, 1794. Whitney brought them and stayed three months to help Miller assemble and install them. Miller oversaw the details of the ginneries through the fall ginning season to the end of the year. He communicated to his ginnery managers through letters, directing them on the construction of "running gear," the selection of waterwheels for the water-powered ginneries, and the purchase of horses for the animal-powered sites. On his insistence, they built ginneries twenty by thirty feet with "six windows of glass."[36] He designed bagging that was woven in a New Haven textile factory and sent his managers directions on how to sew it into bags and pack them with fiber.[37] Miller hired two mechanics, Obediah Crawford and George Chatfield, to build animal- and water-powered mills and install and repair the gins.[38] Interspersed among directives to managers and mechanics were letters denying merchants the right to lease the gin, letters apologizing to planters or cotton factors for ginning or shipping delays, and letters requesting loans. Business, in other words, was good.

Two countervailing events occurred in 1795 that were beyond Miller's control. The first accelerated the adoption of the wire-toothed gin by improving it; the second retarded adoption by castigating it. In January 1794 a ginnery manager hinted to Miller that Augusta mechanics were making toothed gins and that merchant ginners and planters were buying and using them.[39] Miller retaliated, anxious to defend Whitney's patent rights and to protect the profits due to Whitney and the Greene estate. First he issued "A Caution," dated May 1, 1795, and published it in several Georgia newspapers. Miller for "Miller and Whitney" expressed disappointment that "several attempts have been made, under the pretext of improvements on their machine, to trespass on their rights, and to wrest from them their hard earned privileges." Such attempts forced Miller to invoke the force of the law. He reminded readers that Eli Whitney had obtained a patent on March 14, 1794, that it protected his invention for fourteen years from November 6, 1793, and that on June 21, 1794, he had "conveyed" one half of the interest to Phineas Miller, formalizing the partnership. The gin that the partnership alone had the right to make, use, and sell employed "teeth instead of rollers, to draw the cotton from the seed, and a brush to clear the teeth."[40] Curiously, he did not mention the critical breastwork.

Miller deepened the mystery of gin's origins when he suggested in the last paragraph that there were mechanics "who have hitherto been ignorant of their patent, and who have constructed machines upon principles, the use of which is thereby exclusively secured to the patentees." The passage suggested that mechanics had been making "cotton machines" that used "teeth instead of rollers . . . and a brush to clear the teeth" before Whitney received the patent, and who might not know that Whitney had since patented the principle, removing it from the public domain. Miller promised that those who stopped making or using the gins and "deliver[ed] up the machines so constructed" would be spared a lawsuit.[41]

Neither Miller nor his antagonists in Augusta specified the "improvements" that were made to the wire-toothed gin, but a group of Englishmen in Natchez, on the Mississippi frontier did document one important change. On August 24, 1795, mechanic John Barcley submitted a partially completed cotton gin to his patron Daniel Clark Sr., a prominent cotton planter. Clark convened a committee of planters and an "inspector," possibly associated with a local warehouse, to witness its operation. The report they wrote included a description of the gin and a record of their reactions. It was powered with one horse and contained twenty-one "setts or circles of teeth" and "clean'd 8 [lbs. of seed cotton] in 14 minutes," from which they extrapolated that a full-sized gin would "clean 1000 per day with two horses and three attendants." This was "far beyond anything of the kind" and the committee planned to "promulgate to the world its amazing ability."[42]

Two weeks later Barcley ran the gin for another committee who wrote a report that matched the enthusiasm of the first but added details. The day of the demonstration was "damp and foggy," the secretary wrote, conditions that rendered seed cotton "entirely unfit to pass thro' the rollers of the Gin hitherto in use of this Country." Relying on friction, roller gins operated best under dry conditions, when the fiber to seed attachment was weak; ambient moisture increased the strength of the fiber making ginning both difficult and dangerous. The gin the committee watched used teeth rather than friction to remove the fiber from the seed. Committee members reported on September 11, 1795, that the fiber it turned out was "not injured by cutting, but of an equal length and much neater than what is produced by rollers." Some seeds had passed through the gin along with the fiber, as they had with the roller gin, but they were "in so loose and detached a state"—perhaps even uncrushed—that the committee believed they could be easily removed.[43]

The size of the gin disappointed the committee but its outturn elated it. The main cylinder of the Natchez gin was only fifteen inches wide but it "carr[ied] twenty circles of sheet iron, formed in the manner of ragwheels," or round saws made with coarse teeth. Barcley assured the group that the full-sized model would be twice the width, contain "fifty circles," and that it would turn out "seven hundred and fifty lbs. of Clean Cotton." The committee concluded its report with the declaration that "this invention will prove of infinite utility to this Province, and has very far surpassed our most sanguine expectations."[44]

The two reports documented the substitution of round saws for Whitney's wire teeth. The conservative change made the gin easier and cheaper to make and also more reliable. Inserting the wire teeth into the solid wooden cylinder followed by bending and sharpening them was a risky, time-consuming process. Furthermore, the teeth bent in use, abrading the breastwork and damaging the fiber when they did not fall out.[45] Barcley made the conceptual leap from a wooden cylinder pierced with annular rows of wire teeth to an axle loaded alternately with round saws and spacers. The spacers allowed the gin maker to center and adjust each saw between the spaces of the grate so that they would rotate through the breastwork like the wire teeth. It was a flexibility that the studded cylinder did not allow. Barcley claimed to have invented the toothed principle and received annuities from planters who used it, but it is more likely that he appropriated the principle and changed its implementation. A resident of Natchez, he had traveled to the Carolinas in 1793 and returned to Natchez in July 1795. During that time he could have passed through Savannah or Augusta and seen a wire-toothed gin, perhaps one that inspired Whitney or one that Whitney had made.[46] Although the committee reports did not mention the breastwork, neither did Miller, in his discursive "Caution." It can be assumed that the Natchez mechanics made gins on the toothed principle using saw, not wire, teeth.

Unaware of their activities, Phineas Miller made good his "Caution" and sued William Kennedy in Augusta, but he faced a greater threat than infringement. Complaints about the characteristics of the fiber from the toothed gin threatened to retard adoption of the gin. Miller tried to forestall criticism by publishing testimonials from yarn makers who had used the fiber to advantage. Three appeared in the *Georgia Gazette* on November 7, 1795, from managers of New England textile factories with whom Miller & Whitney had done business. They swore that Whitney's gin did not "injure the staple of the cotton" and that its fiber spun into a good yarn. Elizur Goodrich Jr., the son of the Reverend Elizur Goodrich, Whitney's tutor, notarized the statements, adding personal histories of the manufacturers and assuring their integrity.

But textile makers continued to complain about the quality of the fiber the toothed gin turned out. One charged the gin with tangling the fibers, creating "knots," today called *neps*. Neps are tiny entangled masses of cotton that can become scattered throughout the fiber. If this happens, they can neither be picked out nor carded out by the spinner, so they impair subsequent stages of yarn and fabric manufacture. Pinching the fiber from the seed, the roller gin presented no opportunity for fibers to entangle. Two opportunities existed in the toothed gin, one in front of the breastwork, in the hopper as the teeth moved through the seed cotton, the other behind the breastwork, when the brush removed the fiber from the teeth. The hopper was the more likely location of the nep formation that caused Miller problems. Whitney had explained in the patent how the teeth would put the seed cotton into motion in the hopper and create the seed-roll. As the teeth grazed an ever-tightening roll, they pulled off pieces of fiber, not full lengths, producing lint with greater variation than roller-ginned lint. Adulterated by neps and more variable in length than roller-ginned fiber, the fiber produced by the toothed gin was anathema to textile makers.

When Miller brought the problem of nep formation to his partner's attention, Whitney defended the gin and blamed the cotton. "The cotton which the English Manufacturers complain of, must have been *naturally* bad," he wrote Miller on December 25, 1795. "They are the imperfect seeds which are in the Cotton before it is put into the Machine. They are knots which *nature* has made and not the Gin." Even with the "knots," the cotton was more valuable than "if it had been Ginned with Rollers," Whitney believed. Factory managers "Buel and Mackintosh" told him that it was "totally impossible the Machine should produce these knots." Miller must "convince the *candid*" that the gin was blameless. "Them that *will not believe* may be damn'd," Whitney declared.[47] But Miller failed to convince his factors that the neps were natural occurrences, since they were not found in roller-ginned fiber. In his last letter to Whitney in 1795, he seemed on the verge of despair. There was "no immediate employment for" the last ten gins shipped, he wrote Whitney on December 31. They needed the "Approbation of the English factories for our cotton . . . for the purpose of stemming the strong current of opinion which is setting against us here."[48]

Instead of approval, the partners met reproach. On April 30, 1796, the *Chronicle* editor reprinted a letter from "a respectable merchant in England, largely concerned in the cotton manufactory" that had appeared in a Charleston newspaper. His charges effectively halted Miller's plans. The merchant had "closely examined" samples from one hundred bags of Georgia cotton sent to him by cotton factors in Glasgow, and had found it irreparably damaged. "It appears as if it had undergone some severe operation, so much so that its staple is nearly destroyed," he wrote. "I am apprehensive this cotton may have passed thro' that new invented gin. . . . If so, it would have been much better for the planters had they never seen such a thing among them; as it has reduced the value of their cotton from 2s. 3d. to 18d. or 19d. per pound. The Georgia cotton is of a very fine and soft texture, and will not bear such violence to be done to it." He added in a postscript that "the staple or fibers . . . is near one and a half inches long. That which appears to have been spoiled by the gin is little above one half an inch, and is matted together very much to its prejudice."[49] Neps had been textile makers' first complaint but staple length, the primary indicator of quality, trumped it.

"Everyone is afraid of the cotton," Miller wrote Whitney in 1796. "Not a purchaser in Savannah will pay full price for it. Even the merchants with whom I have made a contract for purchasing, begin to part with their money reluctantly. The trespassers on our right only laugh at our suits, and several of the most active men are now putting up the Roller Gins; and, what is to the last degree vexing, many prefer their cotton to ours."[50] Slow to grasp the implications of what he had documented, Miller grappled with two contradictory phenomena: mechanics were building the toothed gins and merchant ginners were using them even as others shunned them in favor of roller gins. In light of the manufacturers' censure, however, the roller gin seemed the only reasonable alternative, even to Miller. One "possibility of Relief from Ruin, in case of the failure of our machines," he wrote Whitney in March, was to "substitute the most approved form of Roller [gin]" at his water-powered ginneries.[51] Business had not recovered by December, and he wrote Whitney regretting "that so promising an invention has intirely failed."[52]

An unintended consequence of the introduction of the toothed gin, Miller discovered, was roller gin enthusiasm. Whitney's patent had spurred wealthy gin maker Joseph Eve to seek a United States patent for his gin, for which he had unsuccessfully petitioned the Bahamian General Assembly.[53] In 1793 he asked South Carolina cotton planter and statesman Major Pierce Butler to submit his petition to the secretary of state, and in 1794 he shipped the descriptions, drawings, and model to him through Dr. Benjamin Rush, a friend of the Eve family, in Philadelphia.[54] A friend of the Greene family, Major Butler may have mentioned Eve's plan to Miller. The information may have prompted the letter that Miller wrote to Edmund Randolph on June 28, 1794. Formerly attorney general, Randolph had become secretary of state after Thomas Jefferson resigned the office in 1793. Miller asked Randolph to delay processing a forthcoming patent for a cotton gin that might have been Eve's.[55]

If Whitney's gin had prompted Joseph Eve to seek a United States patent, manufacturers' opposition to it prompted him to introduce his gin without patent protection two years later. In the spring of 1796 Thomas Spalding announced in the *Georgia Gazette* that he was empowered as Eve's agent to take orders for gins. They would be shipped from Nassau, the Bahamas, for the price of "50 guineas," and the inventor would send "a competent person" to install the machine. A cotton planter himself, Spalding knew that planters bought gins based on their rated outturn and he included Eve's estimates on smooth- and fuzzy-seed cotton. Planters could expect an outturn of "500 lbs. of clean cotton per day" of smooth-seed, Sea Island cotton, and "from 150 to 200 lbs" of the fuzzy, "green seed cotton."[56] The next month, the *Gazette* published a large, front-page notice announcing Eve's arrival. Now the inventor himself, Spalding wrote, could "answer any inquires respecting his cotton gins" and would supply planters with gins in time to "clean out their present year's crop."[57]

Spalding appended a testimonial to his notices that expressed the essence of the criticism of the toothed gin and set the tone of subsequent roller gin claims. It portrayed the toothed gin as the machine that destroyed the staple and the roller gin as the one that preserved it. The high outturn of the toothed gin, however, forced roller gin advocates to address their gin's low outturn and high labor costs. Notices varied in their wording but generally claimed that the gin being promoted processed both fuzzy- and smooth-seed cotton, did not break the fiber, required less skilled labor or fewer laborers than other roller gins, and turned out more fiber of higher quality. The testimonial, written by J. Waldburger for Spalding, was representative. The rainy season had thwarted his plans to experiment with "the green seed cotton" on "Mr. Eve's Cotton Ginning Machine" on a visit to New Providence in the Bahamas, but the inventor assured him that the gin would turn out "in good weather two hundred weight of clean cotton in the course of a day" without "doing any injury to the staple." He had visited ginneries in Nassau and reported that he "never saw more than two people attending the same, one a grown person and the other a small boy."[58]

Spurred by roller gin enthusiasm, men from the professional and artisanal classes of Augusta and environs published notices of their roller gins throughout 1796 and 1797. In July Robert Watkins, a distinguished lawyer and politician, announced his new "Machine for cleaning cotton by Rollers" in the *Chronicle*.[59] In September a group of "disinterested persons of respectability" judged its merits and published their report.[60] William Longstreet and three other members of Augusta's Mechanics' Association, were "happy to announce" that Watkins's gin was "well calculated to clean with great expedition, without injury to the staple." A barrel gin with sixteen to twenty roller gin heads, it could be supplied by one person provided that the rollers did not "exceed 8 or 10 inches in length."[61]

Within the year, John Currie from Shawfield "near the Shoals of Ogechee," William Longstreet, and John Murray each announced the invention of barrel gins they later patented.[62] Currie sought orders for his self-feeding "Patent Gins," built with "any number rollers that the moving power will work," in the December 10, 1796, issue of the *Chronicle*, and had agents in Charleston, Savannah, and Augusta ready to take orders and sell licenses. An impressive list of planters, officials, mechanics, and merchants, among them William Kennedy, testified to the "simplicity, expedition, and durability" of Longstreet's gin. It seemed "impossible that rollers can be arranged to move with less friction, or in a less space," they wrote, adding that the fiber received a "brushing . . . without any injury to the staple," giving it a "superior appearance."[63] When Watkins accused Longstreet of infringement, John Catlett, former president of the Augusta Association of Mechanics, certified that they were "as distinct from each other in their principles and operation as it is possible for two roller gins to be except their both being worked by coggs."[64]

Phineas Miller had not capitulated in the face of resurgent roller gins, merely shifted strategies, juggling the irony he now fully grasped, that the toothed gin was both vilified and coveted. He no longer solicited testimonials from English or American yarn makers. Instead, Miller marshaled his resources and sued all suspected makers and users of gins that incorporated the toothed-gin principle. In 1795, his lawyers filed the first two cases, which were representative of the twenty-five that they prosecuted. The case against William Kennedy, who was charged with using wire-toothed gins, ended with a verdict for the defense. The second, against mechanic Edward Lyons, for making the gins that Kennedy used, ended in nonsuit. A nonsuit ruling precluded a verdict and required plaintiffs to pay court costs but also allowed them to re-open the case at their discretion. Of the twenty-five recorded suits, prosecuted over fifteen years, Miller incurred fifteen nonsuit rulings. Three of those defendants countersued and won. Two cases were dismissed, five were never prosecuted, and seven were left undecided. Miller actually lost only two of the cases. The three he won reshaped Whitney's personal history, the history of the cotton gin, and interpretations of southern development.

The first case involved Hodgen Holmes, an Augusta mechanic who had followed John Barcley in substituting round saws for the wire-toothed cylinder. Miller had suspected Augusta mechanics of making changes to Whitney's gin to avoid infringement charges before the English manufacturers published their recriminations. However, none of them had substituted saw for wire teeth before 1796. Lyons had pirated the wire-toothed gin.[65] In the spring 1796 Holmes, a furniture maker who owned land and slaves, made the substitution but had the temerity to patent and then name his gin the saw gin. Then–Secretary of State Timothy Pickering awarded the patent on May 12, 1796, the same day that Thomas Spalding advertised Joseph Eve's roller gin in the *Georgia Gazette*. Absorbed as Miller was in stemming censure, suing Kennedy and Lyons, and celebrating his May 31 marriage to Catharine Greene, the patent escaped his notice. When he told Whitney about it a full year later, Miller had already planned an offensive. On February 15, 1797, he informed

Whitney that "the name of the Patentee for the surreptitious Patent I think is Robert Homes." He then directed Whitney to "take the deposition of Goodrich and Stebbins on the subject of ratchet wheels," which would prove that Whitney had invented both wire-toothed and the saw gin.[66]

Miller took the case against Holmes beyond infringement. He used it to test and refine patent law, in particular regarding the legality of a patented improvement of a patent held by another. Were it not allowed, the patents of the roller gin makers of Augusta could be invalid because they copied the roller gin principle and made barely distinguishable changes. The legal difference between the gins, however, lay in the originator of the principle. The inventor of the roller gin was an anonymous ancient and its principle so widespread as to be unpatentable. Roller gin modifications thus infringed no one's rights. According to Miller, patenting any modification of the toothed gin, unfairly and illegally preempted the rights of the identifiable patentee of the gin and the principle.

Anticipating the conflict, the framers of the Patent Act of 1793, which superceded the first 1790 law, agreed. The 1793 law stated that anyone who "discovered an improvement in the principle" of a machine could patent only the improvement, and could not "make, use, or vend" the whole machine. Holmes acted legally in patenting the saw cylinder, but he could not, according to the statute, use it in the gin covered by Whitney's patent. However, the law also prohibited the inventor of the original principle from using the improvement; thus, neither Miller nor Whitney could legally use Holmes's saw cylinder. The section further stipulated that "simply changing the form or proportion of any machine . . . shall not be deemed a discovery." It left it to the courts to determine whether a conservative change like the cylinder substitution was an "improvement in the principle" or merely a change in "the form or the proportions." If the court determined that the patent for the improvement was indeed "obtained surreptitiously," meaning that it was a change legally claimed and covered in the original patent, and that the patentee "was not the true inventor," then the court could repeal the "illegally obtained" patent and assess damages in the amount of court costs.[67] According to the statute, Miller's lawyers needed to establish Whitney's right to claim originality and priority for both the wire- and the saw-toothed cylinder. In a court of law, "originality" meant that the idea for both cylinder types had originated with Whitney. "Priority" under the patent law meant, among other things, that Whitney had been the first to develop the ideas but had not previously patented them. Any ruling that affirmed Whitney's claims to originality and priority of both cylinder types would nullify Holmes's patent.

Miller waited over two years before he sued Hodgen Holmes, during which time he used the ironies of adoption to his advantage, maligning the saw gin even as he retrofitted his own gins with Holmes's cylinders. Complaints about staple damage continued to the end of the century, but Miller now blamed it on the saw-toothed gin. His gins, made "with a nice thorn tooth made of wire," were harmless.[68] Still, Miller shipped gins with saw teeth at the customer's request, and he advised a ginnery manager to make the change at the partnership's expense since "we consider the use of saw teeth cut on plates of iron, as completely within our patent right, as wire teeth."[69]

Confident in winning a nullification of Holmes' patent, Miller flouted the law. But neither Holmes nor any other gin maker sued him or challenged Whitney's claims. The patent law allowed a patentee to file an interference if there was reason to suspect that someone else had filed a patent for the same invention. Without incurring the expense of a jury trial, the patentee could appeal to an arbitrator, who would determine the merits of the complaint.[70] The gin makers of Augusta were aware of this option. To a man, Miller's "trespassers" and "intruders" were educated, prominent members of their communities. Numbered among the defendants were two U.S. senators, one Georgia state senator, four grand jurors, a physician, various state and county officials, merchants, and well-heeled mechanics. That none retaliated through the courts suggests that they could not prove originality or priority of the toothed principle. That Miller did not avail himself of the remedy signifies more.

Rather than seek a judgment on interference, Phineas Miller planned to have Holmes's patent nullified. By law he could seek this redress within three years of the award of a patent "surreptitiously obtained."[71] In May 1796 Holmes received the patent; in February 1799 Miller directed his attorney to initiate a suit against him.[72] The delay gave Miller's mechanics time to gain expertise in making the gin saws, Miller time to gather proof that Whitney had conceived of the idea of saw teeth originally and had covered it in his 1794 patent, and Holmes little time to appeal.

Only five documents remain from the case against Holmes, evidence that Miller established the merits of the suit by the May 1799 deadline. On June 30, 1800, Judge Joseph Clay Jr. issued a Rule Absolute enjoining Holmes to show why the court should not repeal his patent. Miller's lawyers followed up with a Plea on July 1, 1800, ordering him to show cause why the partnership should not proceed against him. Holmes's lawyer answered, filing a Plea on February 8, 1802, stating that Holmes had not infringed Whitney's patent because it specified a cylinder "filled with teeth of iron wire only." It further stated that Holmes's patent specified only that the cylinder be "filled with teeth cut in circular metallic plates," which, he added, "the said Hodgen avers to be an improvement of a very important and material nature." George Woodruff for Miller & Whitney and Charles Harris for Holmes responded with a Demurrer on March 10, 1802, each arguing that the other had insufficient cause to proceed. Seated at Savannah, the Honorable William Stephens issued a Final Judgment in favor of the plaintiffs on November 6, 1802, nullifying Holmes's patent.[73]

Holmes retaliated, publishing a notice in Georgia and South Carolina newspapers ten days later declaring that he "has the only Patent which has ever been issued under the authority of the United States, for the cotton gin worked with circular metallic Plates, commonly called the *Saw Gin*."[74] Unsuccessful in overturning the verdict, he succeeded in January 1803 in thwarting Miller's efforts to sell a license authorizing the manufacture, sale, and use of the gin to the South Carolina legislature.[75] Still, Holmes found himself forced to buy a license from an agent of Miller & Whitney allowing him to make the saw gin. "We now hold his note given for that license," Whitney wrote Stebbins on March 7, 1803, relishing it as a trophy.[76]

Two suits followed the repeal of Holmes's patent, both involving prominent citizens of Georgia and both ending in verdicts for Miller & Whitney. In one case, Miller targeted Arthur Fort, an Augusta lawyer and state senator, and John Powell, a Louisville doctor; in the other, his opponent was Isaiah Carter, a member of the grand jury of Waynesboro.[77] He accused each of using saw gins, and none denied the charge. Their defense rested on the claim that since Eli Whitney had not invented it, they had not infringed his patent. The details of the cases are less important than the decisions the judges wrote. The 1810 ruling in the case against Carter quoted heavily from the decision handed down in the Fort-Powell case in 1807. It affirmed and reinforced a history that Miller had meticulously constructed before his sudden death in 1803, one month short of his fortieth birthday.

Miller had initiated the strategy in 1797 by asking Whitney to depose Stebbins and Goodrich on the "subject of ratchet wheels." Expediency, Whitney asked Stebbins to remember, had forced him to use "a coile of iron wire" because he could not find sheet metal to make cylinder teeth.[78] Whitney asked him to sign a deposition of his own composition that included the statement, "Whitney repeatedly told me that he originally contemplated making a whole row of teeth from one plate or piece of metal."[79] Stebbins, then a judge in New Milford (now Alna), Maine, known for his "inflexible integrity," did not reply.[80] His silence suggested that he well remembered Whitney's letters detailing the struggles he encountered making the toothed cylinder and that he knew that none of Whitney's solutions involved the substitution of round saws. Goodrich also maintained silence, but Whitney testified that Goodrich would recall that using "circular pieces of Metal" was Whitney's "first mode or method of making the annular rows of teeth." Goodrich, along with "other persons of science and respectability," had advised him, Whitney swore, that it was "totally unnecessary" to include the idea of saw teeth because it did not affect "the principle of the invention."[81]

Against this, the judge weighed testimony from Augusta mechanics and merchants that they had made and used toothed gins before 1794 and from English mechanics that they had seen similar machines in textile factories. However, he noted the absence of interference filings from the first group and of verifiable evidence from the second, who described machines with toothed cylinders but none that operated in conjunction with a breastwork or grate. The judge determined that the saw cylinder was merely "a more convenient mode of making" the wire cylinder and had resulted in "no improvement" in the patented toothed principle. Whitney had no motive to conceal the alternative mode of construction, he believed, but even if Whitney had had a motive, the omission did not affect the "merits of his invention." Upholding the 1802 repeal of Holmes's patent, the decision written by Judge William Johnson in 1807 and upheld by Judge William Stephens in 1810 affirmed that the wire-toothed and the saw-toothed gin were, in the eyes of the law, one and the same machine.[82]

Judge Johnson transformed Whitney's personal history by rewriting the history of the gin; in doing so he also rewrote the history of southern development. Green seed cotton was more prolific yet more difficult to separate from its seed, Judge Johnson wrote in his Decision. But for want of "some powerful machine for separating it than formerly known among us, the cultivation of it could never have been made an object." Whitney's gin "so facilitates the preparation of this species for use, that the cultivation of it has suddenly become an object of infinitely greater importance than that of the other species [smooth-seed cotton] can ever be." Furthermore, "The whole interior of the

Southern states were . . . languishing, and its inhabitants emigrating, for want of some objects to engage their attention, and employ their industry, when the invention of this machine at once opened views to them which set the whole country in active motion." Supplying "our sister states" with "raw materials for their manufactories, the bulkiness and quality of the article afford a valuable employment for their shipping." Not only was Whitney's gin a boon to national and regional economies, according to Johnson, it had effectively created those economies.[83]

The passage bore an uncanny resemblance to a history written by Phineas Miller in 1803 to Paul Hamilton, the Comptroller of South Carolina, in a lengthy letter contesting Hamilton's cancellation of the state license. Miller wrote that planters had only introduced "the culture of Green Seed Cotton" in 1792, when Whitney ventured southward, and that, "for the want of a suitable Gin, but a small part of it had been prepared for Market." He acknowledged the roller gin but described it as economically unviable.[84] Collapsing two hundred years of cotton production and roller gin use in North America into the moment when Eli Whitney invented the toothed gin, Phineas Miller and Judge Johnson marked 1794 as a turning point in southern development. Before, southerners languished without an effective gin for short-staple cotton; afterwards, the cotton economy blossomed. Arguing for discontinuity, the idea allowed the visualization of a moment and a machine that separated the colonial past from the new republic. Continuity, however, marked the history of cotton and the gin in America. Continuity would characterize the first two decades of the nineteenth century, as saw-gin and roller gin makers competed for dominance in the expanding short-staple cotton market.

CHAPTER FOUR

The Transition from the Roller to the Saw Gin, 1796–1830

Roller and saw gins competed in the fuzzy-seed cotton market from 1796 until the mid-1820s, when the roller gin was marginalized in the small market for long-staple cotton and the saw gin became the gin of the expanding short-staple cotton market. Many factors paced the shift in ginning technologies. Among them were the cultural associations the gins evoked, their visibility and availability, and their relative ease of use. The roller gin referenced a colonial past that signified continuity within a changing society while the saw gin reified attitudes championed by the heirs of the American Revolution. Furthermore, after the efflorescence of roller gin making in Augusta in the late 1790s, roller gin makers squandered opportunities to promote their gins. Saw gin makers maximized theirs through newspaper notices and city directory listings. They also developed an infrastructure, which enabled them to build better, cheaper saw gins and to distribute them widely. Roller gin makers made no similar innovations, retreating from the short-staple cotton market as saw gin makers advanced. But textile makers also paced the change. Once defenders of roller-ginned fiber, they slowly capitulated in the face of larger quantities of the cheaper, shorter fiber that saw gins turned out. Their demand pushed the expansion of cotton cultivation and the shift to the saw gin.

The choice of ginning technology involved agreement among the constituents of the cotton industry—fiber users, fiber producers, and gin makers. Each group was materially invested in the choice because the gin determined the value of cotton fiber, which in turn determined what the user paid and the producer received, although the gin was not the only determinant. Nature endowed the plant with genetic characteristics that husbandry nurtured and harvesting captured. From planting to picking, it took about six months to produce a cotton crop. A gin could destroy its value in seconds. The roller gin preserved fiber length, but it could damage fiber in other ways. Run too fast, the rollers scorched the fiber, discoloring and weakening it. The saw gin sacrificed fiber length for quantity, turning out large quantities of short, uneven fiber. If quality, defined as fiber length, had been the only factor that constituents valued, then the toothed gin would have been immediately rejected as a travesty. If quantity had been the only consideration, there would have been no transition period, because constituents would have adopted the toothed gin immediately. Neither occurred. Instead, the cotton industry supported both gins until a subtle set of factors converged in favor of the saw gin.

Of those, intangible notions of national identity and modernity were the most difficult to document but possibly the most important factors. The rhetoric that tinged discussions of economic development during the early national period militated against the use of a colonial technology, however effective, in the service of a crop poised to enrich the new republic. It worked in favor of the adoption of the saw gin, which, in form and function, represented a rejection of the past and an embrace of the modern. Tench Coxe expressed the new attitude. In the 1780s, he had advocated the expansion of cotton production when roller gins alone were used. In his 1814 *Statement of the Arts and Manufactures,* a report based on the 1810 census, he congratulated planters for transforming "a petty object in little fields" into "an article of extensive cultivation," but he ignored the roller gin. *"The invaluable saw gin,"* Coxe wrote, "was invented by a citizen of the United States" and had "set loose those immense powers of agriculture to produce cotton wool, that were before declined."[1]

Coxe championed the saw gin as the uniquely American invention that it was, and credited it, as Judge Johnson had three years earlier, with transforming the southern economy. At the same time, he implicitly devalued the roller gin because of its colonial and foreign associations. Coxe used the collective and personal pronoun to emphasize the point that Americans had, "by our own invention," contributed to the development of the textile industry with a machine *"to facilitate the manufacture of a staple production of our soil,"* which could be powered with "water, steam, cattle, or hand."[2] The new nationalism invoked by Coxe urged planters to expand cotton production for the sake of national prosperity. It influenced their choice of gin type and, in turn, gin makers in their choice of which to make.

Coupled to nationalism and modernity as factors in the shift of ginning technologies may have been the price of enslaved African laborers on which planters and merchant ginners relied. In the United States in the first decades of the nineteenth century, as in the Bahamas in the last decade of the eighteenth, planters understood that roller gins required the labor of young, strong men, commodified in the market as "prime hands." Animal and inanimate sources powered barrel and Eve gins, and planters and merchants used children to operate them. Although they lowered the cost of labor, neither gin type had lessened the popularity of the foot gin. Its operating cost was tied to the price of labor, either a purchased or hired enslaved African, but also to the price the fiber received on the market. Planters thus considered labor costs along with cotton prices in their choice of gin. Slave prices are incomplete for the early nineteenth century and so variable between markets as to render unhelpful any calculation of relative labor costs pertaining to gin choice. But if the value of enslaved labor is measured in terms of opportunity cost—by the value of the crop they could produce were they not operating a roller gin— then the rationale for planters' decisions might become clearer.

United States planters received higher cotton prices during the transition period than they would for the entire first half of the nineteenth century. In 1795, they fetched 36.5 cents per pound of short-staple cotton; in 1805 the average weighted price at New Orleans was 23.3 cents per pound. It reached nearly thirty cents in 1817, when planters received, on average, 29.8 cents on the pound. During the early twenties prices fluctuated wildly but settled into a downward slope in the late 1820s, hitting 8.4 cents per pound in 1830.[3] Without comparative purchase and hire prices for "prime hands," cotton prices alone would seem a significant factor in planters' gin choices.[4] Increasing cotton prices had motivated Bahamian planters in the mid-1790s to shift to Eve's gin and use their "prime hands" to grow cotton. Prices peaked when United States planters faced the choice of the roller or the saw gin and may have similarly motivated them to shift to the less labor-intensive and more productive toothed gin.

These supply-side factors were never independent of demand-side forces. Responding to consumer demand, textile makers, as the purchasers of fiber, dominated the demand side of the market equation shaping planters' decisions, from cotton production to gin selection. If some were satisfied with roller-ginned fiber from colonies in the Caribbean and South America at the end of the eighteenth century, other textile makers feared that industry expansion would outpace supply and precipitate a crisis.[5] Napoleon's declaration of war against Great Britain in February 1793 and his subsequent interference with British trade in the Mediterranean and the Atlantic disturbed the equilibrium and increased industry tensions. Fustian and muslin makers who had already been searching for a greater variety of staple turned to producers in the United States. Michael M. Edwards, historian of this period of the British cotton industry, calculated that by 1802 the United States "had become the largest supplier of cotton to the British market."[6] Complaints about bags of roller-ginned fiber from South America adulterated with seed fragments overshadowed those about the length and neppiness of United States saw-ginned cotton.[7] Still, in 1802, Phineas Miller competed with roller ginnery managers for the business of upcountry planters.[8]

Cotton gin makers were a crucial, although ill-defined, constituency in the textile industry during this transition period, but they were essential in bringing about the shift of ginning technologies. They appreciated both the planters' and the textile makers' perspectives and understood that their role in the industry was to mediate the contradictory needs of these dependent yet often antagonistic groups. As a malleable idea in 1795 in Natchez and 1796 in Augusta, the saw gin provided them with opportunities to test that skill. It also allowed them to affiliate with the modern through the emerging cotton economy, a burgeoning arena open to clever mechanics. They acquired new technical knowledge within communities of mechanics where they honed it as they experimented with saw gin design. Although saw gin makers availed themselves of newspapers to promote their wares more often than roller gin makers, few identified the type of gin they made. Charting the shift in technology, therefore, necessitates the use of inferential as well as empirical evidence. The conclusion of this undertaking is that several processes occurred simultaneously: the roller gin persisted as the saw gin grew in popularity and gin makers formed communities in cotton-growing regions, laying the foundations for an industry.

Two of the earliest communities of saw gin makers formed during the last years of the eighteenth century, one in Augusta, Georgia, and the other in Natchez, Mississippi Territory. Ironically, none was established in New Haven, Connecticut, where Eli Whitney built the first toothed gin in 1794. In the business of making gins for only four years, he shifted to musket production in 1798, and made gin models only as needed for his legal defense. By then his business associate Phineas Miller had hired mechanics to make gins in Georgia. With funds from Catharine Greene's estate, Miller funded the development of the toothed gin, promoted it, defended it, and hired southerners to make it. The agent of diffusion for the toothed gin, Miller also created the first community of toothed-gin makers.

The introduction of the toothed gin in Augusta sparked the formation of three groups of gin makers. One group rejected the new technology and mobilized resources to introduce more productive barrel gins. The other two embraced the toothed principle. The smaller toothed-gin group worked with Phineas Miller; the larger organized in spite of him. From Miller's perspective, Hodgen Holmes was the most notorious member of the latter because he had patented the saw cylinder. Holmes, however, was only one of five mechanics whom Miller sued for making either the wire- or the saw-toothed gin. Defended by prestigious attorneys Robert Watkins, the same who patented a barrel gin in 1796, and Charles Harris, these mechanics tapped resources that unified them—not only against a perceived outsider but also within their community.[9] The Augusta Association of Mechanics, founded in 1794, reinforced and validated their identities as mechanics even as it may have disapproved of their surreptitious gin-making activities. Apart from Holmes, who purchased a license to build the gin from an agent of Miller & Whitney, none advertised the sale or repair of gins or identified himself as a gin maker.

Mechanics working for Miller benefited from his extensive network of friends and associates. Hired in 1794, George Chatfield and Obadiah Crawford worked throughout Georgia and South Carolina, installing and repairing wire-toothed gins and retrofitting them with saw cylinders after 1798. Crawford stayed with Miller for five years before he left to start his own business in 1799. He purchased a license to make, use, and sell the toothed gin from Miller, who congratulated him on his decision. Miller clearly admired Crawford, whom he recommended highly in a 1798 letter to Stateburg, South Carolina, cotton planter John Mayrant for his sobriety, modesty, and industry.[10] Business trumped affection, though, and he charged Crawford the full four hundred dollars for the license. Miller magnanimously extended him credit "so as to enable you to pay out of the proceeds of the Machine."[11]

Crawford patented his own saw gin in 1807, the year Whitney's patent expired, and was still in business in 1815 when he placed a notice in the *Augusta Chronicle*. Displaying appropriate deference, Crawford addressed the modernizing planters of the upper Savannah River valley, "respectfully informing them . . . that he is prepared for making and repairing Saw Gins, and from a long acquaintance with the business . . . he flatters himself with the pleasing hope of participating with others, a share of the patronage of the Agriculturists of the community." For custom-made gins, he charged "three Dollars per Saw, of 9 or 10 inches diameter"; for a ready-made, thirty-six–saw gin with ten-inch diameter saws, one hundred and twenty dollars.[12]

As the cotton industry disputed the utility of the saw gin, mechanics like Crawford acknowledged its advantages and established manufacturing standards that persisted through the antebellum period. The earliest saw gin makers defined the gin by the number of saws it contained and priced the gin accordingly. They set the range of saw diameters but varied the breastwork and adjusted the gin price to reflect "improvements" or "new plans." No descriptions exist of the saw gins that Miller & Whitney sold, but James Miller, a gin agent and possibly a brother of Phineas Miller, reported a price of $250 in 1806, when Whitney's patent was still in force.[13] After the patent expired, saw gin prices dropped. In 1811 Thomas & J. Reid, an Augusta gin-making partnership charged three dollars per saw, which came to $120 for a standard forty-saw gin. In March 1811, the Reids may have pioneered the production of "steel hardened breasting," a more durable breastwork or grate created either in the forging process or through lamination.[14] Isaac Anthony Jr. may have been a member of this dynamic community of mechanics. In August 1811 he offered for sale "cotton gins . . . On a new and much approved Plan."[15] Since his notice included none of the words that signified the saw gin, such as "saw" or "grate," it is unclear just what type of gin he sold. By the end of the decade, gin makers in central Georgia overshadowed this pioneering community; not until the late 1840s would another reemerge in Augusta.

In 1795, mechanics in Natchez replaced the wire-toothed cylinder with the saw cylinder, and development accelerated sharply. With the support of cotton planters who had buyers for toothed-ginned fiber, they formed a dynamic, long-lived community, far from Phineas Miller's litigious reach. In 1795 Massachusetts-born mechanic David Greenleaf joined the Natchez community and built a saw gin for a prominent planter. Gun maker James McIntyre applied iron casting and carpentry expertise to build a saw gin for Natchez merchant George Cochran for $231. The community's most notorious member, Bennet Truly, built and ran a public gin on the plantation of Ebenezer Rees in 1796 before being accused of gin fraud and of deserting his family. The following year Daniel Clark Sr., Barcley's patron, hired an unnamed African-American mechanic to build a gin. The young English traveler Francis Baily during a visit to Natchez in 1797 reported seeing "several jennies [gins] erected in the neighborhood, in order to extricate the seed from the cotton" and noted that one was "worked by two horses" and turned out "500 lbs of clear cotton in a day."[16] This saw gin–making activity was unmatched in the late eighteenth century. During the same period, Whitney made and shipped wire-toothed gins to Miller in Savannah, who retrofitted them with saw cylinders only on demand.

The Natchez community attracted another Massachusetts-born mechanic in 1807, Eleazer Carver of Bridgewater. Carver had ventured south to build mills but found making cotton gins more profitable. He described the experience in an 1853 article published in a prestigious mechanics' journal. There were no "foundries for the metallic parts of the gin," he remembered. Gin makers cut saws from "inferior sheet-iron" and blacksmiths "forged" them from the bar, before finishing them with "cold-chisels and files." They made the breastwork in the same way. These standard shop-floor practices explain why the saw gin was a considerable investment. As mechanics rationalized and mechanized certain manufacturing processes, the cost of the saw gin dropped, availability increased, and the shift accelerated.[17]

Carver stayed in Natchez during the War of 1812, building sawmills when cotton production and exports were interrupted. After the war, possibly in 1817, he opened the "Temple of Industry," a gin and mill factory funded in part by Seth and Abram Washburn Sr. of Bridgewater.[18] He had come south as a millwright but left in 1819 as a ginwright, his identity shaped by two interdependent communities.[19] Natchez gin makers had refined his woodworking and metalworking skills and taught him how to apply them to the task of building the complex saw gin; planters had exposed him to the slave-based cotton economy in which they were used. By cultivating both groups, Carver gained unprecedented access to what would become the largest market for cotton, and by association for the cotton gin, in the South.

The Augusta and Natchez gin makers established the early foundations of the saw gin industry, but mechanics working in other areas were ambiguous about the kinds of gin they made, signaling both continuity and change. In the first decade of the nineteenth century, Charleston city directories listed four men who called themselves "cotton gin makers." Peter J. Bron in 1801 appears to have been in the business full time, since he did not list another occupation. Benjamin Rigalle and James Symonds hyphenated "cotton gin maker" to other occupations. Rigalle was also a "razor grinder" and Symonds a "mill maker."[20] That the three men lived near each other on "Meeting Street" reinforces the theory that machine makers, particularly in the agricultural South, settled in neighborhoods where they could benefit socially and economically from association with other artisans. This was a short-lived association; none of the three advertised in 1809 when Thomas Fullilove worked at "King-st road," the only "cotton gin-maker" in the Charleston directory.[21] Although Rigalle worked with metal in his razor-grinding business, it is probable that the four men made either foot- or Eve roller gins, supplying coastal cotton planters with the same kind of machines that were used on fuzzy- and smooth-seed cotton during the late colonial period.

Benjamin Prescott, a blacksmith, may have made saw gins. He announced from his shop in upcountry Columbia, South Carolina, in 1808 that he was continuing "the Smith's [blacksmithing] business" on "Mr. Malachi Howell's Bluff plantation." In addition, he added, he "will make and repair cotton gins, and warranted them to be equal to any that are made in the state."[22] His association with metalworking, coupled with his location amidst the short-staple cotton culture, may indicate saw gin production but the date also suggests the roller gin, documented in use in adjacent Sumter County in 1810.

Roller gin maker Joseph Eve added his cachet when he relocated from the Bahamas to Charleston sometime between the fall of 1803 and the winter of 1805. Minutes of the meetings of the vestrymen of St. Matthew Church in Nassau place him in the Bahamas through October 1803.[23] A newspaper notice places him in Charleston in January 1805, when Benjamin Bethel, one of his workmen, reported the theft of a tool chest from the gin shop. In April of the same year, Eve announced that he had relocated to a new shop at Cochran's Ship Yard on the Cooper River near Charleston. He kept two agents in Charleston and one in Savannah to take orders for his "improved Roller Cotton Machine." He also sold licenses to build the gin for "ten guineas" and offered ginning and moting services for "six cents per pound." On September 15, fire destroyed much of his shop and machinery but spared the ginnery. Eve stayed in business by contracting "cabinet makers, and others qualified" to make gins for him.[24] An elite who never listed an occupational title, Eve was respected as a gin maker and his gins were purchased as status symbols. Planter Thomas Hall, for example, solicited no customers but shared his enthusiasm for his new gin equipped "with two pairs of rollers, originally constructed by "Mr. Eves of New-Providence," in Charleston's *City Gazette and Daily Advertiser.*[25] *The Port folio,* a Philadelphia gentleman's journal, reinforced Hall's pride of association and enhanced Eve's status when it featured an engraving of the gin and a copy of the short description of his 1803 U.S. patent in an 1811 issue.[26]

In spite of the prestige the gin connoted, Eve stopped making it after he relocated his family to Augusta and licensed others to make them in the 1810s. Benjamin Bethel went into business for himself in 1810. He had "served seven years to the making of cotton gins," he announced in a Charleston newspaper, and was the "only person in this state" authorized "to build Eve's Patent Gins." He was a symbol of continuity but also an agent of change: in addition to Eve's gin, Bethel made the "common" foot gins as well as saw gins and "impelling powers for the same."[27] Instead of making gins, Eve followed his father's trade and built a black powder manufactory that possibly supplied the federal government during the War of 1812. He also wrote poetry. In 1823 he published a volume entitled *Better to Be: A Poem in Six Books.* Written in iambic pentameter, it was Eve's answer to Hamlet's rhetorical question, "To be, or not to be?" Two years later he patented a steam engine and boiler in the United States and Britain. "Fortune" favored him with a son, Joseph Adams Eve, and a nephew, Paul Fitzsimmons Eve, both of whom became distinguished doctors in Augusta and founders of the Medical College of Georgia.[28]

The waning popularity of the roller gin on short-staple cotton may have been a factor in Eve's decision to change occupations. In 1810, federal census marshals counted cotton gins in some districts, later counties, of South Carolina, documenting the gradual shift to the saw gin. In coastal Beaufort District encompassing smooth-seed, Sea Island cotton regions, marshals listed 2,741 roller gins and 351 saw gins. They found "8 cotton gins," likely rollers, in Marion, a district that spanned coastal and upland regions, and 230 saw gins and "50 sets roler gins" in upcountry Sumter District, an exclusively fuzzy-seed cotton region. In adjacent Kershaw District, however, they found eighty saw gins and no roller gins.

As did saw gin makers, marshals used the number of saws to identify gins. The gins they counted in Kershaw District ranged in size from 16 saws to 140.

The average carried 40 saws. The largest had 140 saws, two had 130 saws, and one had 100 saws. These large gins were dramatic examples of the state of the art of saw gin–making at an early stage of the industry. They also represented a significant capital investment by planters and merchants. At a cost of three dollars per saw, the average-sized gin cost about $120. Assessing the value of the installed base by multiplying the cost of a saw by the total number of saws (3,330) yields an investment in 1810 dollars of $9,990 for the gins alone.[29] The figure excludes the cost of gin buildings, labor, bagging, and associated equipment like scales and wagons.

Marshals did not survey the gins in upcountry Fairfield District just west of Kershaw but they located William McCreight, a saw gin maker and textile manufacturer, in nearby Winnsborough. McCreight himself provided the few extant details of his life in newspaper notices. In 1806 he announced that after "seven years experience in making Cotton Saw Gins," he believed that he had brought them "to great perfection." Since Eli Whitney's patent was in effect and Phineas Miller's agents were active in the area, McCreight quite likely had purchased a license that authorized him make and sell saw gins. He sold them for three dollars per saw if customers picked them up in Winnsborough and for fifty cents more if delivered. In 1809 he also operated a textile factory from which he sold "Homespun cotton bagging."[30] McCreight's manufactory may have failed, as did the South Carolina Homespun Company, a textile factory in Charleston, from the post–War of 1812 economy. McCreight appeared to concentrate on gin and mill building thereafter. In 1817 he placed a notice in Charleston's *Southern Patriot and Commercial Advertiser* announcing the arrival of English "sheet iron" for his saw gins, which he sold on 1806 terms. In the 1830s, he turned to invention. In 1836 he and his son James together patented a "reverse motion" gristmill and an "improved" saw gin, which was manufactured by themselves in Winnsborough, as well as by Bloomfield & Elliott in Raymond, Mississippi.[31]

William McCreight used a labor force composed of white, enslaved, and free black mechanics, a practice that became standard in southern shops. According to the 1810 census, he owned ten enslaved Africans, some of whom undoubtedly worked in the gin shop. He also employed an enslaved man named April. April's owner, possibly also his father, was planter William Ellison, who had apprenticed him to McCreight in 1802.[32] Over the next fourteen years, April learned from McCreight the arts of gin making and blacksmithing and also how to read, write, and calculate. When Ellison manumitted him in 1816, April legally assumed his owner's name and opened a gin shop in Stateburg (Sumter District). As William Ellison, he advertised "Cotton Saw Gins" locally in the 1830s and attracted customers like the Mayrant and Kinlock families, who a generation earlier had patronized Phineas Miller.[33] Later a cotton planter and owner of enslaved Africans, Ellison amassed an estate sufficient to buffer his family and firm from the disruptions of the Civil War. In 1874 the R. G. Dun & Co. credit agency ranked him "hon[orable] upright," "steady hardworking and industrious," and "worth from 7 to 8m$ [$7,000–8,000]."[34]

During the transition period, South Carolina mechanics continued to juggle gin making and repairing with millwrighting and other artisanal occupations. Starting in 1816, William Atkinson and John Workman combined the manufacture and repair of saw gins with the production of other agricultural tools in their shop "at the sign of the Sheaf, Rake & Hoe" in Camden. In an April 1824 issue of the *Southern Chronicle and Camden Aegis,* Workman announced the ongoing demonstration of his "double breasted 40 Saw Gin," which he claimed ginned "100 wt. seed cotton in 7 minutes." The notice suggested that the partners were presenting a fast, new model.[35] In business in 1841, Workman encouraged planters to bring their gins in for repair before the onset of the ginning season so that he might "accommodate all and disappoint none."[36] Like Workman at the beginning of his career, John Graham of Williamburg District, just southwest of Marion District, combined construction of saw gins with another vocation in 1820, in his case the manufacture of "riding chairs."[37]

A woman joined the community of Charleston gin makers in the 1820s. Jane Litel identified herself as a widow and "saw gin maker" in an 1822 city directory, the combination suggesting that she had inherited a gin shop upon her husband's death. Two years later, James Litel, her son or brother-in-law, replaced her in the listing and operated the business through 1837.[38] Litel's brief tenure as the head of a machine-making firm and her use of the title "saw gin maker" in so public a medium, suggests that her involvement in the shop was more than nominal. One of three named female gin makers identified in the antebellum period, Litel highlights the heterogeneity of the southern gin industry.[39] It incorporated men like the elite Joseph Eve and the razor grinder Benjamin Rigalle, wealthy African Americans like William Ellison née April and anonymous enslaved mechanics. It appears even to have admitted gender diversity. Heterogeneity did not convey parity but was mediated by social institutions that established boundaries that were far from porous. Yet diversity in the southern shop was evident from the inception of the industry and persisted through the antebellum period. It stood in contrast to northern gin factories, which exclusively employed New England–born white men.

Burgeoning cotton production attracted a community of saw gin makers to central Georgia during the late 1810s and 1820s, illustrating the accelerating pace of saw gin adoption. Three gin makers advertised out of the region in 1818, two of whom worked in Mill Haven in Putnam County and one in Milledgeville, then the state capital at the fall line of the Oconee River. Martin Thornton of Mill Haven promoted his gin, equipped with "steel plated breast irons," while Elisha Reid of Putnam County claimed that his method of "steeling and facing the breast," increased the longevity of his gins. When A. J. Brown took over Reid's business, he promoted the technique.[40] The gin makers of central Georgia appeared to continue the experiments with breastwork durability advertised by Thomas and J. Reid, possibly relatives of Elisha, in Augusta in 1811. They also struggled with terminology. Thornton, for example, used the term "irons" to refer to ribs, the discrete elements of the breastwork, although gin makers also used the term collectively to refer to all of the metal parts of the gin. Over time, "steel plated" and "faced" breasts became standardized as meaning a flat iron rib reinforced with welded or riveted sheet steel. The term would again change when the cast rib replaced the composite rib later in the antebellum period.

The central Georgia community dissipated like the Augusta community, with some men moving west and others possibly changing occupations. Census marshals compiling data for the 1820 census found only Elisha Reid. Twenty-four years old, he owned thirteen enslaved Africans and appeared to use none of them in agriculture. Instead, Reid dedicated his life, and possibly those of his slaves, to making saw gins. With his brother Templeton, he had established the partnership of "E. & T. Reid" and moved to Columbus, Georgia's westernmost fall line city, in the 1840s. He lost his shop to a fire that ravaged Columbus in 1846 but identified himself as a "mechanic" in the 1850 population census, suggesting that he stayed in the business.[41] The owner of three enslaved Africans at that time, Reid may have worked independently or on contract for E. T. Taylor & Company, a large cotton gin factory that was established in late antebellum Columbus. Not much more is known about Reid, but he is significant because he was a small but full-time gin maker who entered the industry at the end of transition period and helped usher it into maturity.

The 1820 census captured gin production details that reveal a growing gin industry. Census marshals located two gin makers, Robert Freeman and James Nickerson of Jones County, Georgia, who operated shops at the fall line of the Ocmulgee River. They were not water-powered, however. The gin makers reported that they used "shears" to cut breastwork and saws, a technique reminiscent of Eleazer Carver's description of 1807 Natchez.[42] Freeman identified himself as a "gin maker and farmer." He reported that he used six hundred pounds of iron and five thousand feet of pine and poplar in his shop annually, and made about six gins with a total value of six hundred dollars. Although he owned forty-two enslaved Africans, he reported only one male "hand" in his shop. With the same amount of iron but less than half the wood, Nickerson and two hands made approximately ten gins valued at one thousand dollars. The descriptions and high cost of the gins leave no doubt that the men made saw gins.

Other descriptions are ambiguous but suggest continuity. Thomas Johnson of Washington County, just east of Putnam and Jones Counties in Georgia, used six hundred pounds of iron and unreported amounts of steel and wood in his shops.[43] A carpenter, he made "cotton gins and gigs" in his "turner shop" equipped with a "laith," and in his "blacksmith shop." With an annual value of one thousand dollars, the items were in "great demand," the marshal noted in the 1820 census.[44] John O. Grant worked due west at Hawkinsville on the Ocmulgee River in Pulaski County. That he had invested only twenty dollars in his shop yet owned nine enslaved Africans might suggest that he combined manufacturing with agriculture. He used both iron and wood and reported making five gins valued at four hundred dollars. The one man he employed likely operated his one "turning lathe."[45] From the census data, the type of gins that Johnson and Grant made cannot be determined with certainty. Both men referenced a lathe, a woodworking tool used to make roller gins, not saw gins. Additionally, the relatively low cost of Grant's output suggested roller gin production.

Captured by the manufacturing census of 1820 because the value of their annual production exceeded five hundred dollars, Thomas Johnson and John Grant appear to represent the persistence of roller gin makers. Associatively they also represented cotton growers who were reluctant to jettison a familiar technology and cotton users who continued to support the waning technology, even as prices escalated. Yet the saw gin encroached on the roller gin, adopted by new planters who had no roller gin traditions to shed. In Franklin

County, for example, in the northeast corner of Georgia, marshals counted eighteen saw gins in 1820, averaging forty saws each. In Jackson County, southwest of Franklin County, they found sixteen. None reported roller gins.

Cotton planting preceded statehood in Alabama and ginners and gin makers followed, propelling the adoption of the saw gin and the formation of the gin industry. Merchant ginner David Moore opened a ginnery in or before 1816 in Huntsville. He announced that his "old Gin is now in operation" in his "New Gin-House" and promised planters that they would get "the highest price for cotton." Sounding like a cautious roller ginner, Moore stipulated that he accepted only "clean and dry" cotton.[46] The timing and location, however, suggest that he ran an old saw gin, not an old roller gin. Northern Alabama ginners, nurturing no colonial traditions, would more likely have invested in the more expensive saw gin for its ease of use. Planters attracted merchants like Moore, who in turn attracted mechanics with skills that complemented the cotton economy.

Mechanics like Brittain Huckaby were typical. He followed the westward tide of migration settling in towns and cities with industrial districts that attracted peripatetic mechanics like himself. For some part of 1805 he lived in Washington, Georgia, where Phineas Miller had established his "Upton Creek Ginnery Number 4" in 1794. The seat of Wilkes, the most populous county in the state, Washington was a transportation hub. Five roads converged there, one of them the Federal Road. In 1806 he advertised the sale of property, possibly his own, in Greene County in an Augusta newspaper. The property included a "double-geared grist mill," a distillery with two stills, an "an excellent saw-gin, with sixty saws, that is worked by water," along with a "good dwelling house."[47] Huckaby next lived in Milledgeville, the state capital from 1807 to 1868, perhaps traveling the Milledgeville-Augusta Stage Road. If the Greene County offer suggests that he had once owned and operated the conventional cluster of merchant machines, then an 1815 Milledgeville newspaper notice suggests that he had retrenched. He was still a merchant, but he now sold "Dutch fans" for winnowing wheat and "spinning machines," with which, he promised, "a single hand . . . can spin from five to six lbs. of cotton in a day."[48]

In 1824 Huckaby no longer used or sold machinery; he made it. He might have backtracked to Augusta from Milledgeville and picked up the Federal Road, which would have taken him diagonally across Georgia and into northern Alabama and thence to Huntsville, where he next made his home. From Huntsville he advertised his inventions as far as west as Natchez, Mississippi, anxious, it seems, to make up for time and fortunes lost as a merchant. His self-confidence bristled merchants but undoubtedly intrigued planters who were adopting the new ginning technology but were mindful of its problems. "After much labor and expense," he began an 1824 newspaper notice, he had effected "a valuable improvement in the common Saw Cotton Gin, and also one equally valuable in the Corn Mill." The improvement, which he patented in 1824, was for "running gear." He cast both the spur and bevel cogwheels in iron, rather than make them from wood, claiming that the substitution decreased friction and power consumption. He also equipped his gins with "cast steel saws and ribs of a composition metal." The local merchant and millwright David Williamson promptly accused Huckaby of stealing the idea for casting gears from him and informed the public that he had "taken measures to prevent the award of" Huckaby's patent. But Huckaby was more interested in promoting his saw gin than squabbling over patent rights and ignored Williamson's threats.[49]

Huckaby followed convention by experimenting with rib construction but he bucked it by experimenting with flue design. The flue was an elongated box-like structure attached to the rear of the gin, through which the fiber traveled from the brush to the lint room, a temporary storage chamber usually below the gin. By replacing the solid wood component with open wire screening, he eliminated, as he stated, "the necessity of the Cotton Whipper." A whipper was a small, hand-cranked implement that an enslaved African laborer loaded with fiber and agitated, shaking out dirt and leaf fragments. Both short- and long-staple cotton underwent the process, which added cost but also value by producing a cleaner sample of fiber for graders. In theory, the screened flue eliminated the cost of whipping because heavy contaminants dropped through before the fiber reached the lint room. With his cast-iron gearing, cast-steel saws, composition ribs, and a screened flue, Huckaby guaranteed that his fifty-saw gin would "clean 5000 pounds of Seed Cotton in twelve hours, all with a three horse power at the extent," and produce "perfectly clean" fiber ready for baling.[50]

It is doubtful that Huckaby's gin actually turned out 1,667 pounds of lint, one third the amount of seed cotton he claimed it could process daily, but his embrace of the technology and his rigorous experimentation signaled the diffusion of the saw gin within communities of mechanics. His migrations illustrate the relative ease with which individuals moved geographically and occupationally. He traveled along the networks of roads that marbled the southern landscape. They linked county seats and state capitals as well as communities of mechanics and were critical conveyers of individuals and ideas. Huckaby's metamorphosis from merchant to manufacturer illustrates the occupational fluidity typical of the period, exemplifying a trajectory that other gin makers followed.

Natural resources and market opportunities attracted other gin and machine makers to northern Alabama, where they moved saw gin development and the gin industry forward. The 1820 census located three gin makers in Moulton, the seat of Lawrence County, less than twenty miles from Huntsville. The three also made mills and employed nine laborers between them. A. W. Bell was a saw gin maker and screw cutter who worked in Tuscumbia, a town on the southern bank of the Tennessee River, just south of Florence. In 1826 he placed a notice in the local newspaper announcing his recently enlarged business. His gins were "of a quality equal to any now in use" because, he wrote, he had secured a supply of "Cast-steel Saws, which he will finish with double steel breasts or single." "Iron Saws" were still available, he assured less wealthy customers, and they would be "finished in the best workmanlike manner."[51]

Bell's notice marked two processes unfolding concurrently in the gin industry. One had been ongoing, namely the development of a standardized gin. Now cast-steel saws and the double or reinforced grate (breastwork) were routinely installed on the best gins. The availability of gin parts was a new development. As late as 1820 gin makers made the saws and grates they installed in their gins. Bell purchased his, presumably fitting them in frames of his own construction. Although the gin industry never adopted the armory practice of interchangeable parts, the supply and use of gin parts moved the gin industry toward the idea of a standard gin. It had other consequences. The use of purchased parts lowered a barrier to the gin-making industry. Mechanics like Bell could make saw gins without acquiring extensive metalworking skills. This resulted in an increase in the quality of gins, not in a decline, as might be assumed. Instead of cutting saws and grates out of sheet metal, gin makers bought foundry-finished "irons" instead. They produced a more finished, reliable product in less time and sold it more cheaply.

Gin makers also addressed frame design in an attempt to simplify the use and repair of the gin. No saw gins have survived from the late eighteenth or early nineteenth century. The few patents restored after the Patent Office fire of 1836 that addressed the gin's structure described it as an impregnable box that presented a physical barrier to the user. If the restored patents accurately depict the gin, then they deviated from Eli Whitney's original plan. He had designed a hopper with a "movable part," which he directed the ginner to "draw back" in order to clear the hopper of seeds and resupply it with seed cotton. Perhaps gin makers followed the directive but still built gins that impeded access to the saws themselves. Two patents that William McCreight and his son James received in 1836 indicate that gin makers faced just these problems.

Apparently, the McCreights needed to convince fellow gin makers that the top of the gin should be hinged to facilitate ginning and gin repair. As it was, according to the patentees, the ginner was forced to disassemble the gin to fix even a jam in the roll box; in the process, the saws "were liable to injury" and ginning "much more delay[ed]." Furthermore, the nicked and worn ribs "let false seed & trash thro' them," contaminating the fiber. Conventional repair required that "all [of the ribs be] taken out and laid [dressed] by a mechanic." This too necessitated disassembling the "gin head." The solution lay in hinging the hopper and installing "sliding," not fixed, ribs. The changes allowed the ginner "to get at the saw[s] and relieve them when chocked . . . with little labor & without any risk of injuring the saws" and to replace nicked ribs easily.

A cotton gin in the collection of Old Salem in Winston-Salem, North Carolina, appears to have been built according to the McCreight design, if not by the McCreights themselves. It is a forty-saw gin with straight ribs screwed directly into the frame. As in the patent drawing that accompanied the restored second patent, a flat piece of wood covers both the top and bottom of the ribs. The front end of the Old Salem gin hinges open on a doweled pivot located at the base of the hopper, again as in the patent illustration. In the open position, both saw and brush cylinders are exposed, eliminating the need for disassembly. Other features of the gin point to its being a rare specimen. The brush cylinder is a solid cylinder, to which the gin maker attached six flanges. Later brush cylinders were large composite structures with many flanges to create a strong draft. Both the saw and brush cylinders run in leather-lined slots chiseled out of a transverse beam rather than in oil boxes, which supplied constant lubricant. All fasteners are wooden dowels of varying diameters and coarse-cut square-headed screws. The gin's somewhat battered exterior conceals a pristine interior that supports an accompanying note from the donor that it was never used.[52]

Other gin makers made changes that further decreased the cost of making and repairing the saw gin. Henry Clark received a patent twenty days after the

McCreights that offered a plan to lower the cost of grate replacement. In business as Clark & Albertson, the predecessor of the Albertson & Douglass Machine Company of New London, Connecticut, Clark could not claim the invention of a two-part rib. Georgia and Alabama gin makers had advertised "steeled and faced" and "double steel" breasts from the 1810s.[53] The two-part rib, which made up the breastwork, was time-consuming to make. Gin makers typically welded a steel face to a wrought iron base, necessitating the replacement of the entire component when it was worn. Clark patented a procedure for riveting the parts together so that only the more expensive "hardened steel" face need be replaced, not the iron foundation.[54] Edwin Keith, a mechanic at the Bates, Hyde & [gin making] Company in Bridgewater, Massachusetts, followed one week later with a patent covering the hardening of the facing metal by "chilling."[55] Experimenting with assembly and durability, both gin makers addressed the expense of the gin. Clark wanted to lower the cost of replacement while Keith hoped to eliminate the need for a replacement. The first idea never materialized, but the technique of chilled ribs became standard manufacturing practice.

Newspaper notices, city directory listings, and census tallies track the ascendance of saw gin makers and the saw gin, but the body of patent evidence reveals a continued commitment to roller gin technology. From the evidence of restored patents, thirty-five United States cotton gin patents were awarded to thirty patentees before 1830, nine (30%) of whom were also gin makers who had advertised or appeared in the censuses. Of the patents, eleven (31%) were for roller gins, nine (26%) for saw gins. Fifteen (43%) did not specify gin type. Roller gin patentees kept pace with their saw gin peers during the last decade of the transition period, each group receiving four patents during the 1820s. The patent evidence also revealed a geographically broad interest in gin development. Of the twenty-three whose place of residence is known, thirteen (56%) lived in the South; Georgia is represented by six patentees, South Carolina by three, and North Carolina, Alabama, Mississippi, and Kentucky by one each. Ten (43%) lived in northeastern states, New York leading with four, and Massachusetts and Pennsylvania each with three patentees. One patentee claimed British residence at the time of filing. The patent evidence mirrors industry ambivalence over the saw gin, as roller gin makers struggled to remain competitive through patented improvements.

The establishment of three large saw gin manufactories, two in Massachusetts and one in Alabama, signaled the end of the transition period. The factory owners established distribution networks throughout the cotton South that brought gins to planters and gin parts to mechanics. They used the vehicle of the newspaper to promote their newest gins, inevitably describing them as "new and improved" in order to attract customers. The company gin men also shaped the communities of small independent makers, but in different ways. The general availability of their gins suppressed the formation of communities of gin makers in some counties but the availability of ready-made parts facilitated it elsewhere. Everywhere in the cotton South, New England gin makers exerted an influence far greater than their numbers. Southern gin makers outnumbered and outproduced them, but they left fewer records for posterity.

Because of Eleazer Carver's foresight, Bridgewater, Massachusetts, became the site of three large gin-making factories in the antebellum period. They were significant not only because of the gins they sold in the South but also because of the innovations that their mechanics patented and incorporated in the gins. Two of the firms, Carver, Washburn, and Company, and Braintree Manufactory, were established in the transition period. The Bridgewater gins quickly gained brand-name recognition and set benchmarks by which planters and southern gin makers alike measured all gins. Both companies were as long-lived as their gins, many of which can be found in museum collections. Carver parlayed the natural resources of Bridgewater, an iron-rich colonial manufacturing center, with "the knowledge [he] acquired . . . while at the South," into a successful cotton gin manufacturing firm.[56] In 1819 he established Carver, Washburn & Company, which may have been an affiliate of the "Temple of Industry," the gin factory he built in Natchez in 1817. In 1826 he incorporated it as the Bridgewater Cotton Gin Manufacturing Company "for the purpose of Manufacturing Cotton Gins and other manufacturing purposes."[57] Operating under the style of Carver, Washburn & Company (hereafter Carver, Washburn), Carver marketed the gin as the "Carver Gin."

The directors of Carver, Washburn brought a range of resources, affiliations, and skills. They contributed capital but also worked in the shops as mechanics, machinists, and foundrymen. Some traveled south as company agents, selling and servicing gins. Others legally witnessed Carver's patents and patented their own gins and presses. Before joining Carver, Washburn in 1820, Artemas Hale, the company president, had worked as an agent for the Bridgewater Iron Manufacturing Company, a vital institution in the Bridgewater community of mechanics. Bridgewater Iron was staffed from 1797 by members of some of

Bridgewater's pioneer families and had become a local école polytechnic where mechanics were trained; many of them later established prestigious firms or assumed managerial positions in others. Hale was also a lawyer and represented Carver in patent infringement suits in the 1830s. In 1832 he received a patent for a cotton press. Abram Washburn Sr. was another member of Bridgewater's pioneer families who had trained as a blacksmith at Bridgewater Iron before joining Carver, Washburn in 1819. In addition to bringing his son in as a director in 1826, he rode the southern circuit as a sales agent.

The Braintree Manufactory had an equally moneyed and talented directorship, although it was a much smaller factory than Carver's. Ezra Hyde, Jacob Perkins, Increase Robinson, and Nathan Lazell Jr. incorporated the firm in 1823 "for the purpose of manufacturing Cotton Gins, Ironworks, and Machinery."[58] Perkins had also worked as a millwright at Bridgewater Iron. With Nathan Lazell Jr., he incorporated Bridgewater Iron as Lazell, Perkins and Company in 1825.[59] The interlinked directorates deepened the resources available to both firms and reinforced social ties. As principals of the Braintree Manufactory, they made and marketed the "Braintree Gin."

The formidable array of capital and talent allowed the Bridgewater firms to make major inroads into the cotton South by nurturing the planter and gin maker markets. Eleazer Carver had appreciated the possibilities before he left Natchez for Bridgewater. In the spring of 1819, he placed a notice in Natchez's local weekly, the *Mississippi State Gazette,* alerting customers that he had left "a quantity of Cotton Gin Irons and a few Gin Stands" with Rutherford, Fiske & M'Neill, commission merchants whom he had contracted as sales agents. He also used the services of the "Messrs. Smith & Lesslie," local gin wrights who now occupied his "Temple of Industry."[60] They sold his "Irons at their shop" and made "Gin-Stands, Cylinder[s] &c."[61] The merchants sold his "gin irons" to gin makers and his "stands" to planters while his gin makers supplied both. The two-pronged strategy allowed Carver to use his manufacturing resources fully.

The size of the gins that Carver's agents sold confirms that the gins listed by the census marshals in Kershaw District, South Carolina, in 1820 were not anomalies. In August 1820, John Winslow, Carver agent and lawyer, advertised saw cylinders with "62 thirteen inch circles, of the first quality; And a whole Stand will be sold at unprecedented low price of 320 dollars!" At about five dollars per saw—at a time when the average price was three dollars—the gin carried an "unprecedented" high price tag. Winslow repeated the notice in every issue of the *Gazette* into 1821 and may never have found a buyer for his expensive gin.[62]

The Natchez community attracted the partnership of Williams & Cuddy in 1821 and the proprietorship of Zephaniah Burt in 1822, both of which did "gin and mill" work.[63] It continued to pull Bridgewater mechanics like Martin L. Thomas, who arrived in the spring of 1821, and John Edson, who in October published an endorsement from Eleazer Carver.[64] Thomas was a Braintree Manufactory agent and gin maker who worked out of Carver's "Temple of Industry, Gin and Mill Building."[65] He stayed in Natchez for five years, sporadically advertising new shipments of Braintree gins and parts and the sale of his own, unbranded gins, which he offered to "put in operation on reasonable terms."[66] In 1826 he moved to Port Gibson, the seat of Claiborne County on the Natchez Trace. Built with federal appropriations, the Trace was a thoroughfare that snaked 444 miles across Mississippi from Natchez northwest to Nashville, Tennessee, clipping the northwest corner of Alabama. From his new location, Thomas "respectfully" informed cotton planters that he would "construct Gins and Mills, either single or together."[67] The cotton gin agents that flocked to Port Gibson may have pushed him back to Natchez. In 1827 either he or a namesake formed the partnership of Tobey, Thomas & Company there and operated a "Gin and Mill-Wright Business."[68]

That these mechanics could afford advertising marks them as prosperous, but none could compete with Eleazer Carver, whose sales agents blanketed newspapers with notices. These were stationary agents who accepted orders and payments and distributed gins. Twenty-two agents in five states have been identified within his extensive distribution network. In Mississippi, Carver kept agents in Natchez, Port Gibson, Vicksburg in the west of the state; Aberdeen and Columbus in the east; and Woodville in the southwest corner. Others worked out of New Orleans, Vadalia, and St. Francisville, Louisiana; Huntsville, Montgomery, and Mobile, Alabama; Fayetteville and Memphis, Tennessee; and Macon, Georgia. In contrast, the Braintree Manufactory had five agents, all of them in Mississippi. The Triana Gin Factory in northeast Alabama, the only southern factory to compete with the Bridgewater firms for the southwest market during the transition period, contracted six merchant houses in two states. Five were in Mississippi, in Natchez, Port Gibson, Vicksburg, Yazoo City, and Warrenton; one was in Huntsville, Alabama, just north of their factory.

Traveling agents were mechanics based in Bridgewater who traveled south to install, repair, and retrofit gins and to impart a personal touch to an otherwise impersonal market transaction. Company directors or mechanics, the agents bridged the miles conveying Carver's personal interest and returning with compliments as well as complaints. The contact gave the agents the kinds of empirical insights that enabled them to make subtle structural changes in gins in their Bridgewater shops, hundreds of miles from the site of use. But it also translated into pleased customers who published unsolicited testimonials. No gin generated more of these than Carver's. In 1821 the editor of the *American Farmer* published an excerpt of a letter submitted by an Alabama planter. The planter praised "Carver's patent Gins" for having incorporated improvements that "separated the cotton from the seed without cutting or breaking the fibres," allowing Alabama planters to command higher prices than planters in "Louisiana and Mississippi."[69] Readers cared less about how the gin alleviated the problem of damaged fiber than about the higher price the planter fetched. Thus the planter glossed over the details yet still compelled attention. Testimonials like these stimulated sales and earned recognition and respect for ginwrights.

North Alabama gin maker William A. Aikin, owner of the third major gin manufactory of the transition period, capitalized on Carver's visibility. In business from 1808, Aikin moved from South Carolina and in 1819 established a gin factory in Triana, a port town on the Tennessee River. His Port-Gibson agent announced in 1821 that he could make "Cotton Gins of any description . . . on short notice," which he would "sell low for cash." His Natchez agent followed with a notice that suggested that Aikin had made an unpatented improvement that increased outturn. He wrote, "Akin's superior Cotton Gins, comprising any number of saws . . . will gin in the best order from 5 to 6,000 wt [seed cotton] per day." Since the "inventor" intended "to bring them into general use," the agent offered them now "at a very reduced price." Doubts about their "utility" would be "satisfied on examination" and buyers would receive "credit until 1st of January next."[70] Lacking details, the notice nonetheless portrayed Aikin as an innovative gin maker interested in selling fast gins to customers on favorable terms.[71]

In 1824 Aikin himself announced the formation of a partnership with his brother, Cyrus S. Aikin, and took the opportunity to explain his gins and his goals. Having "much enlarged" his Triana factory, he operated W. A. & C. S. Aikin under the new style and continued "the manufacture of Cotton Gins." He reassured customers that the new gins he now made "retained" the "most approved parts of the original plan," although they included "many important alterations and improvements." They were now equipped with "polished cast steel saws and improved double steel breast." Should planters prefer gins on "Carver's model," however, Aikin was able to accommodate having just "procured a lot of Carver's Saws, Ribs, &c."[72]

Unwilling to pander to the popularity of the Carver label, Aikin wanted readers to know that he did not merely assemble gins from ready-made parts. "From his long experience in the business" of making saw gins, he wrote, he had "taken great pains to inspect all the late improvements on Gins in Mississippi and Louisiana" and adopted only the most sound. "Gentleman wishing to purchase" could call on him, and "examine work, prices, &c. and judge for themselves." A flexible manufacturer, he asked customers to specify gin speed, right- or left-hand orientation for the drive band, and their preference for a flue, now also standard.

Like the stationary merchants who handled gins for Aikin and others, the hardware dealers from whom they bought gin parts were essential to the diffusion of the saw gin and the formation of the industry. The Aikins did not sell gin parts but other southern gin makers did. In addition to Smith & Leslie, who made Carver gin parts in the late 1810s in Natchez, hardware merchant Gabriel Winter sold parts from a variety of makers during the early 1820s. In partnership as Winter & Moore and as a sole proprietor, Winter expanded his inventory to include gin irons, saws of varying diameters and tooth density, cylinders, brushes, and "cotton press irons."[73] In north Alabama, David Williamson, who contested Brittain Huckaby's patent right, may have been both a mechanic and a merchant. According to his notices, he kept a "constant supply" of gearing for gins and corn mills in 1824—presumably of his own manufacture—but he also "furnish[ed] gear for Carver's Gins."

The Bridgewater and Triana factories dominated the southwest market, but late in the transition period Samuel Griswold monopolized the Georgia market. Griswold was born in Burlington, Connecticut, in 1790 and moved to Clinton, in central Georgia, no later than 1819 when he signed a "Deed of Mortgage" for two tracts of land in Jones County.[74] Like Huckaby, Griswold was a merchant before he became a gin maker. He changed occupations sometime between 1820 and 1823.[75] He reported no one "engaged in manufactures" in the 1820 census but he had established a gin manufactory by 1823, documented in the 1824 "Report of Such Articles Manufactured in the United States . . . with a Schedule of Factories," compiled by the secretary of state.[76] One year later Griswold formed a partnership with a man known only as "Clark" that bridged the two careers.

Doing business as Griswold & Clark, Griswold placed a notice in 1825 informing planters that the partners "manufacture and keep on hand for sale at their shop in Clinton, Jones Co. Cotton Gins with Steel brists." They also repaired gins, he wrote, and "in the best manner and on short notice." As merchants, they sold "cast iron cog wheels," a standard in Georgia by the mid-1820s, as they were in Alabama. In addition they offered "Wheat Fans, . . . Freeborn's improved Cast Iron Ploughs, . . . wrought Iron Ploughs, . . . Carding Machines . . . for family use, . . . Bedsteads, fancy woodseat and kitchen chairs, with a general assortment of furniture."[77] The notice ended with a solicitation for wood turning orders, suggesting that finish carpentry may have been the partners' specialty.[78]

The Griswold & Clark partnership dissolved within the year and Griswold dedicated himself to making saw gins. He "continue[d] to manufacture Cotton Gins in Clinton Jones County," he informed planters in an 1826 notice, and sold them for "two dollars and fifty cents per saw." Like Aikin, he wanted planters to know that he did not merely assemble parts but experimented with alternate designs and chose the best one. Thus he wrote that after having made "several important improvements in their construction," his gins were "superior to those, or to any made in this part of the country."[79] The claim was safe. The gin makers who had advertised out of the region in 1818 were no longer in business and the Morgan County firm of Allen & Allen identified in the 1824 secretary of state's report was inactive.

Griswold may have been the only identified gin maker in the state in the transition period, but he nevertheless contributed to the formation of a community of gin makers. At its core may have been enslaved and free black mechanics. Famous for his team of enslaved African gin makers who made guns for the Confederacy in 1862, Griswold may have incorporated men of African descent in his shop from its inception. Two "free colored" men lived within his household in 1820 in addition to at least two enslaved men of working age, whom he may have included in the gin or carpentry businesses.[80] His location near Macon, one of Georgia's fall line cities, attracted white mechanics, many of whom left to establish their own factories.

One of Griswold's protégés was Israel Fanning Brown from New London, Connecticut. Brown had come south to join his family in Macon, where his father operated a furniture manufactory. He was nineteen when he decided to move a short distance to Clinton in 1828. Several sources reported that Brown worked with Griswold until 1831, when he returned to Macon.[81] He later moved west to Girard (today Phenix City), Alabama, to manage E. T. Taylor's gin factory; he stayed with the company when it moved across the Chatta-hoochee River to Columbus, Georgia, in 1843. When the business changed hands and names sometime in the late 1840s or early 1850s, he became a part-ner in Clemons, Brown & Company. Having established a reputation by the late 1850s, he formed the Brown Cotton Gin Company in New London in the late 1860s. From itinerant to partner then owner, Brown traveled the path of many antebellum gin makers building a career and a fortune.

By the late 1820s, roads and waterways linked a network of saw gin makers, hardware merchants, and sales agents within the southern states and to Bridgewater, Massachusetts, into an industry that pushed the less efficient roller gin out of the short-staple cotton market. British and United States tex-tile makers encouraged the shift when they learned how to spin the short fiber and how to market the low-end fabrics they made from it. They absorbed the cost of quality by exploiting the gains in quantity.[82] The Carver and Braintree firms in Bridgewater, the Aikins, Samuel Griswold, and hundreds of named and unnamed white and enslaved African mechanics in the South ushered the gin industry out of the transition period and into the antebellum period. The industry encompassed northern firms but was southern by demographics and output. Southern gin makers fostered regional industrialism as they clustered into communities in the seats and towns of cotton counties where they manu-factured the indispensable and emblematic saw gin.

CHAPTER FIVE

The Saw Gin Industry, 1830–1865

The cotton gin industry from 1830 to 1865 is a case study in southern industrialization because it channeled resources into a manufacturing industry that complemented the south's agricultural economy. Census and other data from Georgia, Alabama, and Mississippi, the three largest cotton-producing states, as well as from Massachusetts, the only gin-making state outside the south, substantiate the character of the industry. It was reflected in the number of southern firms, the number of southern-born gin makers, and the number of gins made by southerners. Massachusetts gin makers increased their visibility during the period and planters often patronized them over local gin makers. But southern gin makers exploited their location. Closer to their customers, they offered less expensive but no less durable gins, and were on hand to install and repair them. They also supplied mills and other machinery essential to an agricultural economy. As they had during the transition period, they opened shops in county seats, attracted to and attracting machine makers with complementary skills. The convergence created zones of industrialization where mechanics shared ideas and gin makers applied them to gin design and to firearms construction for the Confederacy in 1862. That factory owners used both free and enslaved African gin makers lent the industry its regional characteristic. That northern-born gin makers headed the largest firms compounded the irony of an innovative industry that blurred regional and racial distinctions as it reinforced them.

The evidence of newspaper notices only hints at the movement of saw gin makers into the seats and commercial towns of southern counties in the 1830s as the cotton economy shifted from the upper to the lower South and into the Old Southwest. South Carolina lost its lead to Georgia by 1830 as the largest cotton-producing state. By 1850 Alabama led, with Georgia and Mississippi following in order, while Louisiana flirted with placement among the top three in 1849. South Carolina, however, maintained its fourth-place position through the 1840s and 1850s.[1] Optimizing opportunity, gin makers moved close to their customers and generated rich detail in their newspaper notices, which evidences the maturing of the cotton gin industry within the industrializing zones of an expanding agricultural economy. During the 1830s cotton was the primary factor in that expansion. As a response to increased demand, production accelerated westward migration and capital formation, dominating interregional and foreign trade.[2] The gin industry complemented the cotton economy and demonstrated that predominance in agriculture did not preclude investment in manufacturing.[3]

Charles M. Johnson represents the entrepreneurial mechanic who epitomized southern industry. "Economy Is Wealth," he reminded planters in 1830 before he offered to supply them saw gins for $2.50 per saw "with fidelity and dispatch." Johnson had moved to Mesopotamia, a town near the Black Warrior River that Eutaw, the seat of Greene County, Alabama, would later absorb. His solicitous closure, "Orders thankfully received," suggests the apprehension of a newcomer.[4] William Avery kept an agent in nearby Erie, the current seat, and adopted a familiar tone in his 1830 notice. It read simply that he had a "first rate Forty Saw Cotton Gin, for sale, made by Mr. Avery," and that interested parties could examine it at their leisure at "J. C. Phares'" store.[5]

In the heart of the Black Belt, central Alabama would attract gin makers to other towns, notably Greensborough and Havana, and become a hub of industrial activity. In 1833, Daniel Pratt, who would become the nation's largest gin maker, moved first to Wetumpka in Elmore County before settling in Prattville in Autauga County. Other gin makers settled in and around Selma, the seat of Dallas County, at the same time. The Sturdevant & Hill factory, in business since 1833, advertised in 1834 their steel saw gins for $3.50 per saw and "Iron Saw Gins (with steel breasts) at $2.50 per saw" and warranted its "50 saw Gins . . . to pick 4000 lbs. seed cotton per day."[6] Cicero Broome and Thadeus Mather expanded the shop of "B. Mather" in 1839 and advertised their "cotton-gin manufactory" in Hayneville, the seat of Lowndes County, where they kept on hand an "assortment of fine and coarse Teeth cast-steel Cotton Gins."[7]

The Aikin brothers, meanwhile, prospered in Triana in north Alabama through the 1830s, advertising in Manchester and Vicksburg, Mississippi, where they kept sales agents. In 1836 Cyrus S. Aikin opened an independent firm with other partners in Columbus, at the eastern border of Mississippi, responding to the westward movement of the cotton frontier and the corresponding opportunities for gin makers. Other gin makers opened shops in the western part of Mississippi, notably in Port Gibson, a town on the Natchez Trace, where Benjamin Kuykendall made "Gin Stands" in 1836. He thanked the "planters of Claiborne and adjacent counties, for the liberal patronage heretofore given" and hoped that it would continue at his new location "where he is determined to supply the demand of the country, in his line of business." Due north in Vicksburg, founder William W. Gaines had announced the opening of his "Gin Stand Factory" in 1832, considered by the late John Hebron Moore, a historian of antebellum Mississippi industry, to be the first and most ambitious in the state. Gaines wanted planters to know that he used "the best cast-steel imported plates" in his saw gins but informed manufacturers that he made and sold a variety of equipment such as "cast iron gearing" and "mill spindles."[8] Abram B. Reading continued making gins after he bought the factory in 1834.[9] Vicksburg merchants handled gins made by "Allen & Adams," "Collins," and a "Milldale" factory, in addition to Aikin, Carver, and Braintree gins.

The majority of gin makers worked in small, independent shops like Kuykendall's, but a few large firms produced the majority of gins. In 1833, another large gin factory opened in Massachusetts, the only state outside of the south with a significant number of gin makers. Bates, Hyde, and Company joined Carver, Washburn, located in Bridgewater, and Braintree Manufacturing, in the nearby town of Braintree. Rhode Island–born Turpin Green Atwood, known as "T. G.," settled in Kosciusko, the seat of Attala County in 1837; he would become Mississippi's largest gin maker. In 1843, E. T. Taylor & Company reorganized and moved from Girard, Alabama, to Columbus, Georgia, eventually dwarfing Griswold, near Macon. In the same year, Eleazer Carver incorporated the E. Carver Cotton-Gin Company with a directorate independent of his Carver, Washburn firm.

The 1850 federal census captured the structure and character of the cotton gin industry at its height. Data were collated from four schedules: schedule 1, "Free Inhabitants (population)"; schedule 2, "Slaves"; schedule 4, "Productions of Agriculture" (agriculture); and schedule 5, "Products of Industry" (manufacturing). The manufacturing schedules for Georgia, Alabama, Mississippi, and Massachusetts mapped a socially, regionally, and economically diverse industry. They enumerated only those manufacturers whose "annual product" equaled or exceeded five hundred dollars, a threshold that excluded the majority of gin makers. In order to include them, the free population schedules were combed and the names of people who used the occupational title "gin maker," or a variation of the title, were collected. The procedure led to an inevitable undercounting since mechanics and founders also made gins. Unless other evidence confirmed gin-making activity, they were not counted.

The population and manufacturing schedules provided an estimate of the number of gin makers in 1850 in the four states under consideration. The Georgia manufacturing schedules have not survived, but at least six firms would have been included. In addition to Griswold and Taylor, they were Griswold protégé Orrin Massey in Macon; Joseph Winship in Madison, the seat of Morgan County; the Watson Gin Factory in Starkville, the seat of Lee County; and John D. Hammock's gin factory in Crawfordville, seat of Taliaferro County, described as "extensive" in an 1849 source.[10] Forty-one gin makers were counted among Georgia's free population. In Alabama, census marshals recorded twelve firms in the manufacturing schedule and eighty-five gin makers in the population schedule. Only eleven Mississippi gin makers appeared in the manufacturing schedule, but there were 117 in the population schedule. Mechanics who called themselves gin makers in Georgia, Alabama, and Mississippi totaled 243 in the 1850 census; twenty-nine of them crossed the manufacturing schedule threshold. Only three of the four Bridgewater firms met the manufacturing schedule criteria (Braintree did not). Twenty-eight gin makers were located in the population schedules.

The manufacturing schedules provide an idea of relative output, although they cannot measure industry output since they did not count gins produced by the small firms. A state historian, who used the Georgia manuscripts before they were lost, preserved some of the data. He reported that E. T. Taylor claimed the production of one thousand gins annually, and Griswold, of nine hundred for a total of 1,900 gins. Alabama's twelve firms reported 950 gins. Of Mississippi's eleven factory owners, four did not report the number of gins they made. The remaining seven reported a total of 328 gins, of which Atwood made 275. From the census data, southern production totaled 3,178 gins. Together Carver's firms reported 309 gins and Bates, Hyde 130 for a total of 439 gins produced annually in the North.

The numbers render incomprehensible the complaint lodged by William S. Figures, the editor of the Huntsville, Alabama, *Southern Advocate* who derided his fellow southerners for their dependence on northern manufactured goods. Sectional tensions flared in 1846 with the onset of the war with Mexico and David Wilmot's proviso amending an appropriations bill to fund the war that called for the exclusion of slavery from any territory acquired from Mexico. Southern Whigs amplified their efforts to achieve economic diversity, but their efforts often went unrecognized. In 1849, Figures wrote, *"Let the South learn to live at home!* At present, the North fattens and grows rich upon the South. We depend upon it for our entire supplies. . . . In northern vessels products are carried to market—his cotton is ginned with northern gins."[11] Figures may have meant to criticize southern patronage of northern manufacturers, a sentiment that southern manufacturers like William Gregg would have shared.[12] The comment, however, seems directed at southern manufacturers, gin makers in this case, who had, in fact, supplied southerners with most of their gins.

Demographic data taken from the population schedules confirm that most gin makers not only lived in the South but were born there as well—yet it also revealed reveals a dependence on northerners that Figures had not calculated. Of Georgia's forty-one gin makers, thirty-three were born in the South, twenty-three, or about 70 percent, in Georgia. Sixty-seven, or 79 percent, of Alabama's eighty-five gin makers were southern-born, of which fourteen, or 21 percent of the southern-born group, were born in Alabama itself; two were European. Mississippi had the most diverse population of the three states, with nine European-born gin makers. Of the remaining 108, eighty-nine, or 82 percent, were southern-born, of which only seven, or 8 percent, were born in Mississippi. The northern-born industrialists, however, headed the largest southern gin factories. From Connecticut-born Samuel Griswold to New Hampshire–born Daniel Pratt and Rhode Island–born T. G. Atwood, each owned the largest firms with the greatest output. They also owned the largest number of enslaved Africans of gin makers in their respective states and reported the greatest holdings of personal property and real estate. E. T. Taylor was the exception, having been born in Virginia, but his partner, Israel F. Brown, was born in Connecticut. Each of them, excepting Taylor, who left the industry, and Brown, who left Georgia in the late 1850s, remained committed to the South and its institutions.

Both southern and northern gin makers were dependent on and committed to the cotton economy, but they stood in an inverse relationship to it. Bridgewater gin makers established firms in industrial towns near raw materials and transportation networks. Invariably they incorporated, following the regional practice of non-durable goods manufacturers like textile makers. Save for the rare Irish-born laborer, they hired locally born white mechanics exclusively. The commercial farmers who surrounded them marketed most of their product—their meat, dairy, and truck crops—to nearby cities and towns like Bridgewater. Bridgewater's gin makers conversely sold their product outside of the region to distant cotton growers whom they rarely encountered face to face. Southern gin makers elaborated a distinctively southern model of industrialism. They established proprietary firms and partnerships, even though textile factories incorporated. They staffed their shops and factories with local, northern, and immigrant white mechanics and with free black and enslaved African mechanics. A mirror image of the northern environment, southern gin makers marketed their output locally, whereas planters sold their commercial product, cotton, outside of the region. Only in the South did gin makers and gin users inhabit the same social and economic space and only there did gin makers shape the slave-based cotton culture as it reciprocally shaped them.

The use of enslaved Africans defined the South and the southern gin industry even though the majority of southern gin makers owned no slaves. Names from the population schedules were matched with names in the slave schedules to generate a list of gin makers who also owned slaves. With the fewest gin makers, Georgia ranked at the top of the category, with 49 percent of the total number of the state's gin makers owning enslaved Africans. Only 39 percent of Alabama gin makers owned slaves, while 27 percent of Mississippi gin makers owned them. Overall, out of a total of 243 gin makers tallied from the 1850 census, eighty-five, or 35 percent, owned enslaved Africans. These numbers establish the typicality of southern gin makers within the South but say nothing about how the 35 percent employed the enslaved Africans they owned. In order to determine how the slaves owned by gin makers might have been used in the production of gins, data from the four schedules were corre-lated.

Gin makers who were not also the heads of their households were excluded from the total number of gin makers. These men were generally dependents in either family or non-family households and typically claimed no real or per-sonal estates (exceptions noted). From the total of 243 gin makers, 55 were eliminated, leaving 188 heads of households, of which 84 were slaveholders, owning a total of 649 enslaved Africans. From this number, women and chil-dren, defined as girls and boys thirteen years old and under, were subtracted. Age parameters were determined from the population schedules where the youngest gin maker was thirteen and the oldest was in his late sixties. This adjustment lowered the number of enslaved Africans to 283. Thirty-four gin makers (40%) owned women and children solely; fifty-one owned work-ing-age men, often but not always in addition to women and children.

Cross-tabulating the population and slave data with agricultural census data reveals a complex association between the ownership of enslaved Afri-cans and their possible use in the manufacture of cotton gins. Of the fifty-one gin makers who owned working-age men, twenty-four were also counted in the agricultural census. Samuel Griswold, Daniel Pratt, and T. G. Atwood numbered among them. In addition to their gin factories, Griswold and Pratt managed other manufacturing interests as well as large plantations. Both men also documented their use of enslaved Africans in their gin shops as gin mak-ers. Less is known about Atwood's labor practices. He planted only 125 acres, to Griswold's 600 and Pratt's 400. Whether he used some of the fifteen men he owned in his gin shop cannot be determined without more evidence. The same can be said of Samuel R. Murphy of Eutaw, Alabama, who listed himself as a "gin maker and planter" in the census and city directories. He owned a to-tal of two hundred and fifty acres; of the thirteen Africans he owned, four were working-age men. The case is equally ambiguous for Jesse Watson of Lee County, Georgia, who quite probably used the nine adult men he owned in his gin factory and gristmill in Starkville, rather than in tending the eighteen "improved" acres he reported.

The case of the Avery family of Perry County, Alabama, highlights the difficulty of generalizing, from census data alone, a single pattern of slave use within the gin industry. David Avery farmed extensively but called himself a "Gin Maker" in the 1850 census. His brother or nephew Richard lived as a de-pendent in his household and was also a "Gin Maker." Avery owned six men, one of whom he identified as a "carpenter" in the slave schedule; it can be assumed that he worked in the shop. Bryant Avery, who lived nearby, called himself a "Farmer" but boarded three non-family members, each of whom called himself a "gin maker." That Bryant was in some way engaged in the family gin-making business is further suggested by the fact that, of the fifty-seven slaves he owned, he identified seven as mechanics in the slave schedules. Curiously, all of them were also listed as "mulatto" and "fugitive." From the census data, the southern-born Avery men divided their resources between manufacturing and agriculture, deepening their investment in both between the census years, adding mill making and milling to the operation by 1860.

The case for the use of enslaved Africans in gin shops is strongest for factories that owned slaves and for gin makers who owned men but no land. Two factories, both located in Alabama, were listed as owners of men in the 1850 slave schedules. The Campbell & Heisler factory in Dallas County also owned land, complicating speculation on how the firm used the six men it owned. The Mather & Robinson factory in Lowndes County also owned six men but neither the company nor its partners appear in the agriculture schedules. Undoubtedly, partners Thadeus Mather and E. P. (Enoch Poor) Robinson used the men in their large steam-powered gin factory in Hayneville. Among the gin makers who were not located in the agriculture schedules but owned men were John Du Bois of Greene County, Alabama, who owned five men; E. T. Taylor in Columbus, Georgia, who owned twenty-four; and Erwin Elliott of Carrollton, Mississippi, who owned four men. They were a minority in the gin industry but, with Pratt, Griswold, and Atwood, among the most influential. Their use of enslaved Africans, however speculative, represents the persistence of a pattern begun in the transition period. Although their names are unknown, these enslaved Africans were factors, along with white and free black gin makers, in the industry's development.

The cotton gin industry was dominated in number by small southern proprietorships and partnerships using an array of family, wage, and slave labor. In gin output, it was dominated by four southern firms, each of which was headed by a northerner, and by four incorporated northern firms. The collective history of these firms point to an eclectic industry, at once conservative and innovative, populated by individuals who dedicated their lives to gin design and construction and others who combined gin making with house carpentry and millwrighting, religious ministries, and planting. Samuel Griswold was both eclectic and conservative; he was not only one of the industry's largest manufacturers but also an influential teacher who mentored a generation of gin makers.

"Careful, proud, honest," according to an R. G. Dun & Company report, Samuel Griswold attracted talented mechanics to his Clinton, Georgia, gin factory and trained them along with the enslaved African mechanics he owned.[13] Israel F. Brown, who in the 1840s managed E. T. Taylor's large Columbus factory, had spent three years with Griswold, from 1828 to 1831, when he may have worked with the enslaved and free Africans Griswold included in his household in the 1820 census. Daniel Pratt had worked his way from Temple, New Hampshire, to Clinton, Georgia, arriving at Griswold's shop in 1831. He stayed only two years before setting out for central Alabama, where he established a factory and foundry. Abraham Massey moved his family from Rockingham, North Carolina, to Clinton in the late 1820s and may have begun working with Griswold then. In 1831 his older son Orrin, and possibly his younger son William, joined Pratt in Griswold's shop. Unlike Pratt, though, Orrin Massey stayed ten years before leaving to establish the Massey Gin and Machine Shop in Masseyville, three miles from Macon. Abraham Massey continued working for Griswold as a sales agent through the 1840s.[14] At least six enslaved African men worked in Griswold's shop at the same time. Griswold named them in an 1842 contract that authorized his son-in-law, Francis S. Johnson, to manage the shop. The contract stipulated that "The Negroes. . . . viz., George, Jerry, Henry, Little Jackson, Matthew, and Wilson," work as gin makers, not "waggoners."[15] Griswold held Johnson legally liable if the men were not used as the skilled mechanics they were.

Griswold also interacted with local gin makers. A remark published by his grandson connects him with the Hammacks (also spelled Hamack), a group of three brothers who operated a gin shop in Crawfordville, just northeast of Griswoldville on the Georgia Central railroad line. Southern-born, they operated within a partnership, making and repairing gins in a small shop. Lacking a foundry, they bought the saws and cast parts from a hardware merchant, or possibly from Griswold, and assembled them into frames they manufactured. They owned no slaves in 1860, although they reported five "hands" in the manufacturing census, which represented some combination of family labor and hired white or enslaved African mechanics. Men of modest means, they experienced the dramatic rise in fortune over the decade of the 1850s, as gin makers did elsewhere. Griswold's grandson may have reflected Griswold's own opinion when he recalled the brothers as "very reliable and honorable men who made gins in a small way all [their] life."[16]

The high turnover in shop personnel that Griswold registered in the 1850 and 1860 censuses perhaps illuminates another facet about the man and his shop. Griswold mixed local, northern, and immigrant, as well as enslaved mechanics in his shop. This heterogeneity not only reflects the region's labor market but also the primacy of skill over demographics and ethnicity. Yet few white mechanics stayed. Census marshals found "assistant manufacturer" H. W. Dorsey and "overseer" Thomas H. Stallworth, both Georgia-born, in the gin shop in 1850 but by 1860 Georgian Henry Adkins and Gilson W. Harris, from Maine, had replaced them at the management level. Griswold had added an Irish laborer and two mechanics who were relatives of Harris. Mechanics were a notoriously mobile population and these transitions may exemplify the reality. It contrasted, however, with Prattville, the town that protégé Daniel Pratt established, where gin makers formed relationships that rooted them to place and region.

Of all the gin makers, more is known about Daniel Pratt because he left a substantial body of correspondence, business records, newspaper articles and editorials by and about him, in addition to parlor gins that historical societies now collect and display.[17] Like Eleazer Carver, Pratt was a mechanic who worked his way south from New England. He left New Hampshire sometime in the 1810s and passed through Milledgeville, Georgia, in 1819, where he may have met Elisha Reid. It appears that Pratt did not turn to gin making then but concentrated on honing his carpentry skills building houses and completing a mansion for Samuel Lowther, a neighbor and friend of Samuel Griswold's in the 1820s. He gained experience "building boats" before joining Griswold in 1831.[18] When he left Griswold two years later, his wife, Ester Tichnor, and two enslaved mechanics accompanied him.[19] Pratt built on a deep foundation of technical knowledge when he established his first shop in 1833 at Elmore's Mill, in Wetumpka, the seat of Elmore County. He moved to McNeil's Mill the following year, as he planned the construction of a permanent factory.

At McNeil's Mill, Pratt established a shop culture that attracted and retained talented mechanics. Among the hands who worked in the shop in 1836 was twenty-four-year-old Georgian A. H. Burdine, who would, however, leave to join a thriving gin community in Holly Springs, Mississippi. He held three gin patents by 1860.[20] Burdine worked with William Orr, a saw straightener who was related either to the gin-making Orrs of Talladega or to the nearer Orrs of Orrville in Dallas County. Among the enslaved mechanics Burdine worked with were Dick, whom Pratt bought in 1834, and Daniel, whom he had purchased from a slave dealer in Jones County, in addition to the men who accompanied him on his trek.[21] From their temporary quarters at McNeil's Mill, the gin makers shipped, in 1836 alone, 128 gins ranging in size from thirty to fifty saws.[22] In 1838 New Hampshire–born Amos Smith joined as a partner, and in 1839 the partnership moved into the permanent, water-powered factory located on Autauga Creek, in Prattsville, later contracted to Prattville.

Through the 1840s, Pratt and Smith, with Griswold's oversight, assembled a team that balanced loyalties and fostered unity as they developed the gin factory as the heart of their industrial village. In 1842 they hired mechanics Nathan Warleigh and Western A. Franks and contracted E. P. Robinson, who also ran a gin factory with Thadeus Mather in nearby Hayneville. The partnership added sawmills, gristmills, and flour mills, as well as the regionally renowned Prattville Manufacturing Company, a textile factory. Together they furnished "the Cotton planter with Gin Stands, [and] Cotton Osnaburgs of as good a quality, and as cheap as they can be procured elsewhere." The boast originated with Pratt, who wrote a lengthy article about Prattville at the request of James D. B. De Bow for his *Review* in 1846. Two years later Griswold sent his son Elisha Case, known as "E. C.," to Prattville as a partner and manager.[23] Amos Smith brought in his nephew Ferdinand E. Smith, who joined his son George in the gin shop.

Ferdinand was an inveterate diarist and commemorated his first day at work. On January 1, 1848, he wrote, "Commenced work for D. Pratt & Co., in the Gin Shop at forty dollars per month including board. Boxed one gin of first quality. In the afternoon attended the annual meeting of the Fire Engine Company, had the honor of being elected Hoseman." That was Saturday. On Monday, he hurt his fingers "in a bad fix between the shafting and the end of a plank" and was out for a week.[24] In 1851 George capitulated to Ferdinand's insistence that he too keep a diary. Northern-born and privileged, the two men were atypical southern gin makers but nevertheless captured, from the vantage point of wage laborers, the details of shop floor practice, the use of enslaved African gin makers, and the vicissitudes of the Pratt-Griswold relationship.

Pratt characterized himself in the 1846 *De Bow's Review* article as a gin maker who supplied fast gins at prices planters could afford. His mechanics made "about five hundred" saw gins annually, which were available at prices tiered to quality. "The 1st are made with double breasted large wing brushes calculated for long flues. These I sell at $4 per saw, the 2d are single breasted, breast same as the 1st quality, calculated for long flues, these command $3.50 per saw; the 3d quality is a lighter gin, with a 16 inch breast calculated for short flues-these are sold for $3 per saw. They run lighter, and gin equally as fast as the other, the saws and breasting on the 2d and 3d qualities are same as the first."[25] The details impressed as well as informed planters already versed in the mechanics of ginning. To them, he explained further that they could run their gins fast without lowering the value of their cotton by having the saw teeth dressed so that they "pass through the breast so that the whole length of the tooth will strike parallel with the rib where it passes through." If they would also dry their seed cotton thoroughly, then they could run their gins fast without problems. Pratt warned readers away from the fancy gins that independent gin makers were then touting as solutions to low cotton prices. He encouraged them to buy his "common" gins made "on the most simple plan I can adopt to have them answer the purpose."[26]

The Smiths described the details of the manufacture of the common saw gin. Working in batches or "lots," they first cut up lumber into fronts, top and end boards, hollowed, smoothed, and turned them before "jointing," "gluing up," and "dressing" them "off." Grinding, straightening, and training saws was exceeded in difficulty only by breasting, or installing individual ribs in the roll box or hopper. As the final step in the process, they applied lacquers or paint to preserve and decorate the frame. They then "boxed" the gins for shipment before rolling them down the gangway onto gin wagons. Ferdinand also made casts or patterns for the ribs of the breastwork and for oil boxes in which the ends of the brush and saw cylinders ran. He had the latitude to introduce incremental changes in gin design and the time to make drafting tools for a local architect. He also repaired waterwheels, renovated rooms, and rebuilt furniture. There were noteworthy inclusions about farm work. Pratt's best "breaster," Ferdinand regularly helped neighbors hay and "reap rye" and even helped Pratt kill hogs.[27] Neither owner nor employee worked exclusively in the shop; they combined shop work with farm labor, balancing commercial and subsistence activities.

About thirty white and enslaved African men passed through Pratt's gin factory in the pages of Ferdinand and George's diaries. Georgia-born men like Norman Cameron, John C. Hearn, and Thomas J. Ormsby joined Pratt in the

1840s and stayed with him through 1860. Others like Henry Hunt and Marcus C. Killett (or Kittell) came from North Carolina and also stayed. New Hampshire–born Merrill E. Pratt, a nephew of Pratt's, joined the team in the late 1840s. Ferdinand's brothers Frank (first name Benjamin) and John came in the 1850s. These men worked exclusively in the gin shop. Other gin makers were contractors who worked on their own account in the shop or in their own shops. E. P. Robinson, for example, worked principally out of his Hayneville factory. Typical of nineteenth-century machine shops, the arrangement was sometimes problematic, as when Robinson delivered a batch of cylinders with "heads" turned too small for Ferdinand to use.[28] Other men launched affiliated careers. Ephraim S. Morgan worked as a "gin workman" when Ferdinand Smith arrived in 1848 but later managed Pratt's Sash and Blind Shop, where George Smith sometimes worked. In addition to his cousin, Morgan, and M. E. Pratt, Ferdinand recorded working with Simon Tichnor, Daniel Durden, a Mr. Reese, a Mr. Carr, and "Billy," an enslaved mechanic, during his first year in Prattville.

Many others passed through the shop, their origins rarely recorded, their skin color and labor status barely distinguished by naming conventions. The cousins reserved the honorific "Mr." followed by the family name for authority figures, although Amos Smith was always "Uncle" to Ferdinand and "Father" to George. The cousins called each other by given names, but George shortened Ferdinand to "Ferd." Friends were addressed by their given names in the diaries, as were enslaved mechanics. Ferdinand diligently recorded the names of the men with whom he worked, but his entries suggest an unexpected blurring of social distinctions. Friday, October 31, 1851, passed like most other days for Ferdinand: "At work breasting all day. George has gone to Vernon to fix a gin. Jim has been at work for me all day at dressing the edges of box heads."[29] George was his cousin, who was traveling about 150 miles northwest to the seat of Lamar County, doing the job second in importance to making gins: repairing them. Jim was an enslaved mechanic whom Amos Smith had bought that March who would spend the day doing George's shop work. Ferdinand used no qualifiers but simply given names to distinguish the two men.

Of the nine enslaved African mechanics mentioned by the Smiths in their diaries, Jim was the only one whose personality emerges from the entries. Although they worked with him almost daily for four years, neither cousin mentioned his age, skin color, or other physical characteristics, or even which of the partners owned him. Nevertheless, both were impressed with his abilities, remarking regularly in their diaries, with characteristic understatement, "He gets along pretty well." A seasoned gin maker, Jim got "out high boards and heads to go on the fronts" of gins on his second day in the shop. He also "got the rib boards out and commenced on the heads." On the third day, Ferdinand left his "fronts" for Jim to finish as he "commenced breasting." At mid-month, George remarked that he had "finished off a lot of tail pieces today."[30] Whether a quantity or a compliment, "a lot" became a standard qualifier for Jim's work. On the twentieth of the month, Ferdinand recorded that "Jim has been at work dressing out common gin lumber. He is getting out a lot, enough for sixty gins of fronts, top boards, and rib boards."[31] As Jim dressed lumber, the enslaved mechanic Jackson put saws and spacers on cylinders and breasted gins.[32] In April, Amos Smith "bought a very likely Negro boy named Lawrence," who joined them in the shop.[33]

Work proceeded uneventfully until August, when the blurred lines that separated color and caste sharpened. Problems seem to have begun in early August, when Jim missed five days of work due to a "bowel complaint."[34] He returned on the 11th and worked with George daily on end and seed boards until the 22nd, when he was again absent.[35] According to George, Jim came back on the 25th and dressed seed boards for him. Ferdinand recorded what else transpired that day. A "Mr. Carrol came into the shop this morning to settle with Jim for some misdemeanor Saturday night. He gave him a few sharp stripes, but I fear not enough, for his benefit in the long run." Carrol had whipped Jim in front of his fellow workman, yet Ferdinand felt that the pain and humiliation were insufficient to humble him. With no recourse, Jim returned to his routine, although Ferdinand complained in October that he "get[s] along slow."[36] For the rest of the year Jim worked almost daily with George but was indispensable to Ferdinand, who made gins on his own account and used shop workmen as subcontractors. When he tallied the wages he owed them for 1851, Ferdinand acknowledged "forty days work by Jim and seven by Mr. Barrows."[37] For the next four years, Jim worked regularly, gaining experience and biding his time. On September 5, 1855, he escaped. George noted the loss with the entry: "Jim ran away this evening."[38] Ferdinand simply ignored it.

A dogged work ethic capped the talent and commitment of Ferdinand and Jim, but their expectations were profoundly divergent. For Smith, his tenure at Pratt's manufactory was prelude to a career that culminated with the formation of Smith Sons Gin & Machine Shop in Birmingham, Alabama, in 1886. As Jim, Billy, Lawrence, and Jackson honed their skills, they increased their value to their owners, not their own opportunities for social or economic mobility. Furthermore, as skilled slaves, they were ammunition in the proslavery argument volleyed against abolitionists who charged that slavery degraded the slave. Here was evidence of achievement, not degradation. But the proslavery argument ignored the fact that the enslaved pursued personal achievement in spite of, not because of, slavery. Yet slaveholders, anxious to preserve their property rights, translated the skills that slaves acquired into evidence not only for the success of the institution of slavery but also for its necessity. In 1859, Pratt wrote that slavery was "the only way the African can be improved physically, morally and religiously."[39] For men like Pratt, manumission was unthinkable; self-purchase untenable; self-hire unconscionable. Self-appropriation was the only avenue to social mobility that the enslaved mechanic could pursue, and Jim took it. Enslaved Africans in the gin industry, as in other southern machine-making industries, represented the stark ironies of industrialism in the slave South, where regional and racial distinctions were at once blurred and sharpened, and opportunities enlarged and limited.

For all but the enslaved, gin making represented the means to economic and social mobility. Hardship emanated from lists of the many birthplaces of gin makers' children, recording nearly as many moves, but better jobs that paid higher wages mitigated it. The life history of Pratt gin maker Western A. Franks, outlined from 1850 census data, was typical. Born in North Carolina in 1800, he moved to South Carolina, where he met and married his wife Caroline, four years his junior. They had their first child, Olivia, in Georgia in 1831, but by 1835, they had settled in Alabama, where they had five more children. Having worked for Pratt since 1842, Franks worked as a "millwright" in his own shop in 1850. Charles P. Riggs was typical of another genre of gin maker. Born in Alabama in 1837, he was separated from his parents and lived with a poor farming family in Dallas County, Alabama, in 1850. Since the couple's two children were toddlers, Riggs probably worked the farm with his guardians. He was still a dependant in 1860 but had changed households and become a gin maker. Riggs had moved west to Greensborough, where he lived with the brothers Abraham and Ingraham Rhodes, who were also gin makers. He claimed no wealth but had made an important first step up the ladder of occupational mobility. Franks and Riggs started from different points but traveled the same road to economic independence.

Employment was assured Riggs in Greensborough, the largest town in Greene County with roughly 1,400 inhabitants in 1860, a commercial and industrial center and home to upwardly mobile gin makers. Shumake & Smith, "manufacturers of improved Cotton Gins," published a large "card" in Gayle Snedecor's 1856 county directory, which included a list of the county's gin makers. Retired physician Lyddleton Smith was not named but may have been involved in the partnership. He also co-owned a factory with William B. [or P.] Ellison organized under the style Smith, Ellison & Company.[40] Smith owned no land but five working-age men who undoubtedly worked in either one of his factories. Either simultaneously or sequentially, North Carolina–born Ellison operated the steam-powered factory of Ellison & Company that reported making one hundred gins in 1860.

Robert Johnson worked as a gin maker, either independently or for one of the factories from the late 1840s to 1860. He jumped in wealth from nil in 1850 to fifteen hundred dollars in 1860 when he boarded a gin maker, John Morrison, possibly an employee. Johnson owned no land but a total of three slaves, all of them working-age men. Alabama native John Du Bois was also a respected Methodist minister. He operated a factory with his brother Joseph C. In 1850 he owned "1 lot," valued at one hundred dollars, as well as five adult men who must have worked in the shop. From a personal value of three hundred dollars in 1850 he reached ten thousand dollars in 1860, in part due to the success of a gin innovation he patented in 1850.[41] Among the other gin makers in Greensborough were Jason Oaks, who had previously worked as a carpenter in Blount County, and Patrick H. Sherron, who had left New York at the age of forty to make his fortune in the South.

On average, Mississippi factories were smaller, the communities more dispersed, and the wealth accumulated by individual gin makers more modest than those of gin makers in Alabama and Georgia. T. G. Atwood was the largest, but ranked fourth in wealth after E. T. Taylor in 1850 and third after Griswold and Pratt in 1860. The least known of the big four southern gin makers, Atwood was not unknown to the two leaders. Born in Rhode Island in 1800 or 1801, Atwood followed the tide of northern mechanics who trekked south in the 1810s and found themselves in Samuel Griswold's factory. A property deed places Atwood in Clinton, Georgia, in 1827, the year Pratt married his wife Ester in Macon, and one year before Israel F. Brown arrived from Macon.[42] While Brown left in 1831 and Pratt in 1833, Atwood resided in Clinton in 1836 when his son, David Chase, was born. The burial of his wife Elvira was the unfortunate marker situating him in Kosciusko, the seat of Attala County, Mississippi, on the Natchez Trace, in 1839.[43]

Atwood harnessed the Pearl River to power the Atwood Gin Factory as well as a gristmill and a flour mill, although he would convert to steam power by 1860.[44] Unlike the gregarious Pratt and the avuncular Griswold, Atwood was a private man who avoided publicity. He may have directed his sales agents to place the succinct notices that appeared in Mississippi newspapers advertising his gins. Cozart & Clarke placed one in an 1848 issue of the Aberdeen, Mississippi, *Weekly Independent* that stated simply the availability of "Atwoods Celebrated Gins" and that they could furnish gins of "any size and description." An adjacent discursive notice published by Eleazer Carver's Aberdeen agents dwarfed it. Atwood's brother William C., who may have been with him in Clinton in the 1830s, joined him in Kosciusko in the late 1840s. They appeared as co-owners of "T. G. and W. C. Atwood," a gin factory that was listed in the manufacturing schedule in 1850, and which R. G. Dun & Co., agents described as "large scale." They operated a more diversified, steam-powered factory in Vicksburg and, like founder William Gaines, manufactured "Gin Stands" as well as textile machinery such as "cotton spindles."[45]

The Atwoods' reputation and that of their gins spread throughout the region despite the cryptic notices. In 1851 an R. G. Dun & Co. agent wrote of the brothers that they were "indus[trious] wor[thy?] men and that their gins were "in a high repute here."[46] Mississippi agriculturist and self-appointed gin critic Martin W. Philips thought there was room for improvement. In an 1850 journal article, he wrote "that the Messrs. Attwood can make a stand that I would prefer [to Bridgewater gins], and if they would procure the best timber and try themselves, they would soon secure a name imperishable."[47] The editor of the Vicksburg *Sentinel* disagreed, attesting that the Atwood stand was "carefully and strongly" built.[48] Another critic saw an Atwood gin at a state fair in Jackson in 1859 and remarked that it was a "beautiful piece of workmanship, with all of the latest improvements."[49] Israel F. Brown acknowledged him with a "testimony" to his "excellent gins" in an 1870 Clemons, Brown trade catalog. Private but not parochial, Atwood stayed abreast of changes in saw gin design and incorporated them in his product line.

He also nurtured a gin-making community comprised of white and enslaved African mechanics, sparking a modest industrializing trend in central Mississippi. His family was one of the first to settle in Pleasant Ridge, six miles west of Kosciusko, the seat of Attala County. In 1850 he was the sole gin maker. By 1860 he had hired two Prussian ginwrights, one of whom, James Dodd, was his foreman. He then competed with Scarbrough & Robinson, a gin factory in Kosciusko, for the services of the four independent gin makers who lived nearby. Atwood planted some of his land but owned thousands of unimproved acres. His dual occupation of "Cotton Gin Stand Manufacturer and Farmer" complicates speculation on how he used the fifteen men he owned in 1850 and the twenty-three men he owned in 1860. It would not be unreasonable, however, to assume that he followed Griswold and Pratt in investing in enslaved African mechanics to help him in the gin shop. Like other southerners, they balanced opportunity on cotton production and gin manufacture.

Manufacturing and agricultural opportunities pulled other enterprising mechanics to county seats and commercial towns in Mississippi through the 1840s and 1850s. Their engagement in national trends and technological innovation was evidence enough for the late John Hebron Moore to declare antebellum Mississippians modern. Amasa Davis would have agreed. He operated the Natchez Gin Stand Manufactory from 1841 through 1860, using the city foundry, located across the street from his shop, to make his castings. In 1847, he made a cutting-edge saw gin that incorporated the latest improvements in the saw, grate, and brush. They ranged in size from "50 to 75 saws" and sold for "$3.50 per saw," the same price as the second of Pratt's quality stands, and the Bates, Hyde & Company's Eagle Gin. Not to be outdone by the New Englanders who also supplied gin parts, Davis carried "Saws, grates, boxes, bolts, &c."[50] Yet Davis, born in Maine, did not ignore agricultural opportunities. In 1850 he owned fifteen hundred acres, of which he cultivated six hundred that yielded nineteen bales of cotton. He was excluded from the final list of slaveholding gin makers because he owned one young woman solely and must have hired factory as well as farm laborers.

Mississippi's county seats pulsed with gin-making activity in the late antebellum period. In 1860 Natchez was home to a diverse gin community that included Ruth Sprague, a twenty-six-year-old Rhode Island woman who called herself a "gin maker" in the 1860 census, and twelve men who hailed from Ireland, England, and France, as well as upper Midwestern and mid-Atlantic states. They either worked as independent proprietors or as contractors for John Spitler and Richard Wilds's Natchez Foundry & Gin Factory, which had changed hands since opening in 1836 but continued to manufacture a range of tools and machines. In Vicksburg the Atwoods participated in a community populated by M. Donovan, who ran a foundry and gin factory, and six independent gin makers. Holly Springs, in northern Marshall County, first settled by white farmers in 1830, had become a transportation hub and commercial and industrial center by 1850. In the 1840s J. W. Brooks operated a shop in nearby Wyatt, competing with James H. Taylor's gin factory in Holly Springs.[51] In 1850, ten gin makers worked in the city either for or independent of the Ivey & Cunningham gin factory. Ten years later, the names had changed and the factory reorganized, but ten gin makers, including former Pratt mechanic A. H. Burdine and the Manuel & Cunningham gin factory, supplied local planters with gins.

Two of the most important Mississippi communities were located in Columbus and Aberdeen, the seats of Lowndes and Monroe counties, respectively, both adjacent to the Alabama state line. After Natchez and Vicksburg, Columbus was the third-largest city, with a population of 3,000 in 1850. In 1837, when Cyrus S. Aikin moved from Triana to establish a gin factory there, it was already a transportation hub with steamboat service to Mobile. The *Columbus Democrat* carried notices from Aikin's Triana factory, Carver, Braintree, and less well-known gin makers. In April 1837, New Orleans gin maker Alexander Jones advertised his patented "Double Gins" in its pages. He had the expensive saw gins manufactured in the New York printing press factory of Robert Hoe. Martin T. Collier used its pages to inform readers that he had recently moved from Monroe County, Georgia, to Louisville in Winston County, Mississippi, where he operated an "extensive Cotton Gin Manufactory," but that he planned to relocate to Columbus. Peter McIntyre announced the availability of "spinning machines, with cast iron frames" in addition to "the flying shuttle loom."[52]

The panic of 1837 took its toll in Columbus, as elsewhere, resulting in the failure of Cyrus Aikin's firm, among others. In 1840 Aikin announced the formation of another partnership with founder Thomas W. Brown that may have met a similar fate.[53] Aikin may have returned to Triana, but Brown stayed to nurture a relationship with Tennessee gin maker Leonard Campbell, with whom he established a partnership in 1854. They styled their firm "Campbell & Brown," inverting the customary placement of the financier's name. It may have pushed out Hayden & McCraw, a gin factory that operated during 1853 and 1854, which an R. G. Dun agent thought could not "fail to prosper" because it had "a complete monopoly."[54] In 1856, Campbell & Brown competed with the diversified Wolfington & Spears foundry, which employed Martin Campbell as an inside contractor to make gins in 1856. Sometime in the late

1850s, Leonard Campbell dissolved the partnership with Brown and moved due north to Aberdeen, signaling the growing attractiveness of the seat of Monroe County for enterprising machine makers.

That Aberdeen recovered earlier than Columbus from the aftershocks of the panic of 1837 is evidenced by the influx of gin and machine makers, led by S. S. Ward, who operated his gin shop in Aberdeen in 1844.[55] By the late 1840s the Aberdeen Gin Factory and Machine Shop dominated, competing only with ubiquitous Carver & Company agents and one Atwood agent. It was owned and operated by twenty-two-year-old William T. McCracken from Alabama. North Carolina–born mechanic Benjamin D. Gullett moved to Aberdeen from Eutaw, Alabama, about the same time. The two men quite likely worked together, but neither revealed much about himself. McCracken listed himself only as a "machinist" in the 1850 census, claiming four hundred dollars in assets. Gullett, who would become one of Mississippi's foremost gin makers, listed no occupation, wealth, or land except for five enslaved Africans, one of whom was a working-age man.[56] In February 1851, McCracken added a foundry, renamed the factory the Planters' Machine Works, and sounded the tocsin of sectionalism that other gin makers would chime. "I invite every Planter and all others who feel an interest in a success of a Southern Manufacturing Establishment," he emphasized, "to visit my Machine Works—and see the vast amount of Machinery I have erected for the more speedy and correct manufacture of Cotton-Gins, &c." He was also "ready to build steam engines . . . together with almost every description of Millwork, Machinery for Cotton Mills, Steam Boat Work," as well as "every description of Heavy Iron Turning, Screw Cutting, Drilling and Slatting."

A notice McCracken published in a May 1851 issue of Aberdeen's *Weekly Independent* moderated the heady February advertisement, documenting an episode of labor unrest in the industry. His foundrymen went on strike "for higher wages, or restriction on hours of work." McCracken summarily dismissed them and advertised for two finishers and one pattern maker, to whom he promised "A permanent situation with liberal wages."[57] Apparently, planters had responded with orders to McCracken's February notice, but instead of hiring more workers, McCracken worked those he had harder and refused to compensate them adequately. There is no evidence that the strike prompted McCracken to replace the white mechanics with enslaved African mechanics, but there is also no evidence that he did not supplement his staff with hired enslaved mechanics. McCracken himself owned one slave, a young woman;

his son Thomas B. McCracken helped him in the shop. Local historians debate what happened in the wake of the strike. Some believe that Gullett bought the establishment, but this is doubtful. He never branched out into other machine-building trades but concentrated on saw gin design, pushing it to its limit. Starting in 1856, Gullett managed a succession of factories, first Gullett, Gladney, & Co., then Becket & Tindall, before moving to New Orleans. Leonard Campbell and Peter Reynolds, both unaffiliated gin makers in Aberdeen in 1860, may have worked for one of these innovative firms.

A small community of industrializing Mississippians, composed of machine builders, textile manufacturers, and planters, congregated in Woodville, Wilkinson County, in the southwest corner of the state. A railroad linked Woodville to St. Francisville, Louisiana, a port on the Mississippi River with steamboat service to Baton Rouge and New Orleans. Mississippi and Louisiana state legislators chartered the railroad company in 1831 but the panic of 1837 halted construction.[58] Completion coincided with the economic recovery; an in-migration of manufacturing firms, including a textile mill, followed. In March 1841 Beach & Keller placed a notice in the *Woodville Republican* announcing the availability of *"new gins,"* which were "made to order, and repairing done with despatch." They also manufactured "Running Gear and Presses of every description."[59] Victor N. H. Netterville's effusive notice overshadowed this modest announcement. The Woodville native operated a gin and machine shop where he made gins "warranted to be equal if not superior in point of beauty, utility, and durability, to those manufactured by any other establishment in the United States."[60] Still in business in 1850, Netterville made the same range of machines as Beach & Keller, in his Woodland Cottage Gin Factory. He met the criteria for the manufacturing schedule, reporting a one-thousand-dollar investment in a "horse"-powered plant, staffed with two men, one of whom could have been an adult man he owned, but since Netterville also farmed, how he employed the man could be debated.

James Alexander Ventress snubbed these gin makers and seven others who worked in Woodville from the late 1840s to 1860, patronizing Eleazer Carver and his gins in spite of calls for southern patronage in the wake of the Compromise of 1850. A lawyer, state congressman, and cotton planter, Ventress was also an amateur mechanic. Between 1854 and 1857 he bought three Carver gins from local agents but was dissatisfied with all of them. He wanted one of the fashionable three-axle gins that Gullett had popularized. Since Carver did not make them, he decided to patent one himself. He corresponded with P. B.

Tyler of American Machine Works in Springfield, Massachusetts, in 1856; with Joseph E. Carver, Eleazer's son, in 1858; and with Philadelphia machinist John L. Tuttle from 1857 to 1858. Tuttle had patented a "cylinder for gins and cards" in 1856 and worked with Alfred Jenks & Son on a new gin type called the "cylinder gin," then also fashionable. With Tuttle's help, Ventress received a patent in 1858.[61] By no means the only planter to patent a cotton gin, Ventress nevertheless displayed a remarkable level of knowledge of mechanics. However, it was his loyalty to northern gins and gin makers—one shared by other large planters—that southern gin makers tried to divert in their sectionalist appeals.

Samuel Griswold was convinced that gin price, not prestige, was the pivot on which gin sales to wealthy planters like Ventress hinged. After musing on market share, he explained to Daniel Pratt in a January 1850 letter that Carver took "all the country trading to New Orleans," selling mostly to "the large Planters," gins priced from four to five dollars per saw. He supposed that if the Griswold-Pratt partnership could keep up their prices, they could "continue to get some of Carvers Customers."[62] Yazoo, Mississippi, cotton planter Richard Abbey had said as much three years earlier. He was a "decided enemy" of inexpensive gins and chastised planters who bought them. "A cotton gin requires the most perfect mechanical precision in all its parts," which, he assumed, warranted a high price. Many southern planters preferred to save "fifty or a hundred dollars in the price of a gin stand" but lose hundreds on the value of their crops over the projected five-year lifespan of the gin, a dangerously short-sighted approach, Abbey felt.[63] Like Ventress and Martin W. Philips, Abbey invested in Bridgewater gins in the late 1840s, eschewing Pratt and Griswold's staple, the "common" gin.

Neither the output of the Samuel Griswold and E. T. Taylor factories nor the westward trend of the cotton economy thwarted a spate of gin shop openings in Augusta in the early 1840s, as the depression lifted. William Jones pioneered in 1841 when he listed his "cotton gin manufactory" in the city directory; others followed slowly.[64] The proprietors of one of the largest firms would come from Sparta, the seat of Hancock County, to the west of Augusta, but not until the mid-1850s. Until then, Garrett T. Oglesby, his son Zehas, Jesse Matthews, and Jonathan Gladdin, all Georgia-born, made and repaired gins for local farmers; only Oglesby also farmed and owned slaves. Neighbor Thomas J. Cheely may have urged the senior Oglesby to leave Sparta, or Oglesby may have solicited Cheely's support in a new venture. The two men combined resources to open the McIntosh Gin and Mill Factory in Augusta, named after the street it fronted. In 1854, Cheely announced in the *Southern Business Directory* that "he had secured the services of Mr. G. T. Oglesby" and "established a large Gin Factory on the Canal." He had bought the "Patent Wing Brush," possibly Carver's 1845 brush cylinder patent, and used it in the prize-winning "Oglesby Gin."[65] By 1860, another gin factory had joined Cheely's in Augusta, owned either by Floyd W. Finch, Thomas Wynn, or Samuel C. Reid, the three wealthiest of the city's eleven gin makers.

The War of 1812 and the panics of 1819 and 1837 punctuated United States political life and spurred economic development and industrialization. The threat of secession exerted its own agency as southern manufacturers entreated southern planters for their patronage in the 1850s. But the Civil War imposed radical change. Historians of southern industrialization have studied the mobilization of foundry and forge owners and their enslaved and waged laborers, who not so much retooled as redesigned their product line and accelerated production. They cast and bore cannon and made other heavy weapons for the Confederacy. Southern cotton gin makers retooled. They converted their gin shops to gun shops and supplied the Confederate government with a range of small arms from revolvers to carbines. Record keepers of the caliber of Tredegar's Joseph R. Anderson or Buffalo Forge's William Weaver did not emerge from the population of southern gin makers. Gin makers left only scraps of evidence, which nevertheless indicate the skill levels that gin makers achieved and applied to the production of guns.

Two gin factories were documented as receiving contracts for small arms from the Confederate government. The less well-known is the Davis & Bozeman gin factory in Coosa County in central Alabama, possibly located in the county seat of Rockford. The partnership was comprised of "Gin Maker" Henry J. Davis, whose personal and real estate totaled $13,500, and "Farmer" David W. Bozeman, whose estate exceeded one hundred thousand dollars in 1860. They had invested twenty thousand dollars in a steam-powered factory, which they staffed with eighteen male "hands." The men made two hundred and fifty gins annually in addition to "threshers and fans," easily qualifying for the manufacturing census. How many "hands" were enslaved is difficult to estimate. Davis owned no land but three Africans in 1860, two of them adult men who probably worked in the shop. Planter Bozeman reported twenty-seven slaves, but since he also owned an extensive plantation, it is again difficult to estimate labor allocation. Sometime in 1863, the Georgia natives retooled and successfully bid on a contract to manufacture small arms. From 1863 to 1865 they made approximately nine hundred rifles and eighty-nine carbines, the locks of which they stamped "D & B Ala." with the year of manufacture.[66]

Samuel Griswold, located near the Macon Arsenal in central Georgia, left conclusive evidence that both the white and enslaved African gin makers who worked for him had attained the same high level of machine-making skill. In June 1862 Griswold formed a partnership with Arvin Nye Gunnison, a gin maker who had served as his agent and managed his New Orleans warehouse since 1858.[67] The date of the partnership coincided with the award of a contract to manufacture revolvers. To the editors of *Macon Telegraph,* it signified the "power of the South to supply her own wants. We certainly had no idea that a manufactory of Colt's Pistols would spring up near Macon in 1862."[68] W. C. Hodgkins, superintendent of the Macon Arsenal, was skeptical when he found "twenty two machines running worked by twenty four hands, twenty two of whom are Negro Slaves."[69] When out of a lot of twenty-two Griswold & Gunnison revolvers, only five "were found fit for service" in an October proofing, the skepticism seemed founded.[70] But the gin makers became better gun makers and the revolver became a staple of the Confederate arsenal.[71]

In October of 1864 Griswold wrote Col. James H. Burton, then-superintendent of the Arsenal, expressing his wish to rent out "my Machinery & hire out my negroes & Teams for the ensuing year." The Gunnison & Griswold partnership had increased the number of "negroes" to thirty, whom "good judges" had deemed "fair mechanics." Their "5 or 6 white mechanics" had received conscription orders "from Richmond." When asked if the enslaved machinists could be separated or relocated, Griswold answered, "The negroes are already well quartered. Their families are on the place & it would be preferable to have them remain here."[72] Before Federal troops destroyed the Griswoldville armory in 1864, the enslaved machinists had manufactured between 3,600 and 3,700 "Colt Model 1851 Navy" revolvers.[73] Caught between personal achievement and social morbidity, Griswold's enslaved machinists had stayed at the bench during the war, leaving a legacy fraught with irony: the gins they made facilitated the expansion of the cotton South and the guns they made defended it.

Not only for its incorporation of enslaved gin makers is the cotton gin industry a case study of southern industrialization, but also for its demographics, output, and structure. Gin makers migrated to county seats and commercial towns throughout the South, drawn there by opportunities in manufacturing as well as in agriculture. They in turn attracted other gin makers who, with resident machine makers, created zones of industrialization where ideas were shared and machines built in service to the South's cotton economy. The saw gin was its emblem, but not an inviolable one. The economic crises that had depressed cotton prices and stymied internal improvements had provoked new criticism. Planters complained that the saw gin damaged the fiber, lowering already low prices. At their urging, gin makers scrutinized each component of the saw gin and patented hundreds of conservative changes promised to mediate market swings. Others introduced new types of gins. In the process, gin makers improved the performance of the saw gin, refined their mechanical skills, and broadened the scope of the gin industry.

CHAPTER SIX

Saw Gin Innovation, 1820–1860

The price shocks that the panics of 1819 and 1837 delivered to United States cotton planters reverberated throughout the gin industry. Planters hit by low prices demanded gins that delivered longer, cleaner fiber, which they hoped they could sell for higher prices. At the same time, they knew that British textile makers, still their largest customers, wanted quality fiber at low prices. They looked to gin makers to mediate these mutually exclusive demands. Some gin makers responded conservatively, others radically, but both groups declared as their shared goal nothing less than the perfection of the cotton gin. Conservative mechanics committed to the saw gin patented hundreds of changes to the gin's components, promising gins that minimized fiber damage and maximized outturn. Whether their fancy attachments, or boxed brushes actually worked was less important than what they signified. For planters unwilling to limit production, they represented solutions to low cotton prices. For gin makers committed to the toothed principle, they signified mastery of the machine and the market. Yet the short-staple cotton market rewarded quantity, not quality. Saw gin makers balanced their rhetoric and fancy attachments accordingly, building ever larger, faster gins.

The low cotton prices that accompanied the two major economic depressions of the first half of the nineteenth century did more to prompt change in the cotton gin industry than they did to curb cotton production. The first price shocks came during the panic of 1819. In September 1818, on the eve of the panic, the market returned to planters 32.5 cents per pound, the highest price of the antebellum period for short-staple cotton. In 1825, the weighted average at New Orleans was 11.9 cents per pound; by 1829, it had fallen to 8.9 cents. Production followed an inverse curve over the same period. Short-staple cotton production rose from 261,233 bales in 1818 to 532,915 bales in 1825 to 762,800 bales in 1829. Falling as it did within the framework of the market revolution of Jacksonian America, the panic of 1837 had a major impact on planters and gin makers. In its wake, cotton production vacillated but generally increased, as prices fluttered but followed a general downward course until they bottomed in 1844 at 5.5 cents per pound, the lowest of the antebellum period.[1]

The changes that saw gin makers advertised and patented during the transition period were artifacts of the incremental modification of the saw gin as it transitioned toward adoption, conditioned by the cotton prices that planters received. Both the process of technological evolution and the economic environment influenced what saw gin makers offered planters and how they presented their products. Elisha Reid publicized his introduction of a "great improvement, viz: that of steeling and facing the breast" in 1818, emphasizing two points that would have appealed both to flush planters and those later hit with declining prices. The "improvement" increased the gin's longevity, he emphasized, but more importantly allowed them to "gin infinitely better."[2] Better ginning would yield higher grades from cotton samplers and higher prices from cotton factors, the comparison suggested and planters would have hoped.

By 1824 the saw gin had passed the threshold of adoption, but planters had not recovered their loses. William and Cyrus Aikin, facing a larger cotton economy and even more precipitous declines, engaged the superlative in their newspaper notices. They advertised their gins as "superior" and constructed on a "new plan," offering gins with "Polished Cast Steel Saws and improved Double Steel Breast[s]," in 1824. But in a competitive gin market they also felt pressured to present their credentials and their philosophy of gin making. At a time when mechanics were known for their mobility, they reassured planters that they had "long experience in the business" and that "the establishment

will in future (as formerly) be able to give entire satisfaction." When gin makers increasingly assembled gins from purchased parts, they informed planters
that they had "taken great pains to inspect all the late improvements on Gins
in Mississippi and Louisiana," and incorporated and warranted only those
that were proven reliable.[3] The notices that the Aikins and other gin makers of
the 1820s published were testimony to their engagement in the evolution of
the saw gin and in the mediation of the unpredictable cotton market.

Bridgewater, Massachusetts, gin maker Eleazer Carver participated in the
process and mediated the market through his gins and patents. He had worked
in Natchez from 1807, shaping the fledgling cotton gin industry, which in the
South and the North, contributed to the industrialization that the War of 1812
had spurred. When the bubble of the postwar expansion burst in 1819, Carver
returned to Massachusetts to "improve said gins." From then until 1832, he
built on "the knowledge acquired by me while at the South," and "occupied &
devoted" his "whole time and attention" to making gins.[4] The gins he shipped
southward in the 1820s were but one aspect of his reaction to the market crisis;
the patents he took were another. Carver received at least two patents in the
1820s, both in June 1823, for a "Cotton-gin saw and grate" and a "Cotton-gin."
Since he did not restore either after the 1836 patent office fire, there are no accompanying descriptions or drawings.

The first patent was clearly a modification of gin elements within the hopper, the "saw and grate," but Carver may also have patented changes behind the
grate or breastwork, in the region of the brush cylinder. In 1821 north Alabama
planter J. R. Bedford described as an "invaluable improvement" to "Carver's
patent Gins" an attachment "which operates as a screen and fan, separating
trash and dirt from the cotton, as the wheat fan separates chaff from wheat. Its
principle is simple, and its construction easy." It had been introduced "in the
vicinity of Natchez," according to Bedford, and with it the gin would "separate
the cotton from the seed without cutting or breaking the fibres." Since he did
not specify whether the attachment operated on seed cotton or lint, precisely
where it sat in the gin was unclear. What Bedford clearly stated was the benefit
planters would derive from the patented change. Where once Natchez cotton
was discounted, it now "commanded two to four cents more than Red River
cotton and 3 to 6 and 7 cents more than Tennessee and Alabama."[5] The article
expressed the prevailing sentiment that technology would stem the market decline, yielding planters greater profits and delivering textile makers better fiber.

Other gin makers patented and promoted gins promised to achieve the same end, but Carver alone articulated mediation as the role of the gin maker and perfection of the saw gin as his goal. He ruminated on his life as a gin maker and on the industry's goals in an 1858 patent extension petition in which he discussed his motives and ideals. Like other gin makers of the 1830s, he had faced demands for longer fiber from textile makers who had adopted Richard Robert's self-acting mule, and for larger gins from planters who had introduced longer-stapled, larger-bolled cotton varieties. The demand- and supply-side changes, as well as the economic crises, had forced Carver to make reciprocal "changes in the Saw Cotton Gins." They were necessary, he wrote, "in order that the greatest profit might be realized both by the Planter & Manufacturers of Cotton."[6] Mutual profitability could accrue only if textile makers could buy fiber at a relatively low cost and if planters could sell it for a relatively high price, and only a gin that delivered quality fiber in quantity could, in theory, deliver "greatest profit" to both groups. From 1832 to 1845, Carver delegated gin making and dedicated himself to "experiments" that he believed were "indispensable to perfecting the said Saw Gin."[7]

The goal of "perfecting" the saw gin—that is, designing a machine that would deliver large quantities of high quality fiber and thus mediate industry conflict and blunt price shocks—drove change in the saw gin industry through the 1850s. To those ends, saw gin makers and mechanics, even hobbyists, patented conservative, small changes to each component of the gin. In front of the grate, they reevaluated the hopper, the saws and saw cylinder, and the grate itself. Behind the grate, they rethought the brush cylinder, but more than that, the interaction of the teeth and the brush (fig. 6.1). They did not ignore the framework of the gin, but rather incorporated changes to it within patents for other parts. Gin makers were also concerned with pre-ginning processes such as seed cotton drying and cleaning and post-gin processes such as lint cleaning and pressing; they held patents for seed cotton driers and whippers and lint cleaners. They also patented power sources. Gin makers purchased patents held by machinists in other industries, if the invention furthered their goals. They also patented outside of the gin industry. Carver, for example, patented a "rolling mill for circular saws &c" in 1838, which he used in his shingle business as well as in his cotton gin business. Crossover patents like Carver's complicated the counting of relevant saw gin patents. Regardless of the patentee, therefore, only patents that could be explicitly tied to the saw

Fig. 6.1. Fifty-Saw Gin, 1842 (67½″ wide × 36½″ high × 32″ deep; 519 lbs). The hopper to the right obscures a row of fifty iron saws that revolve through a fancy steel breastwork. The large brush cylinder to the left is a distant cousin of Eli Whitney's four-flanged brush cylinder. It would be covered in operation in order to contain the lint and create sufficient draft. Notice the square profile of the saw cylinder behind the large flywheel, and the circular brush cylinder. Both cylinders sit in inks—cast-iron troughs, visible in the foreground—which held lubricant. Courtesy, Smithsonian Institution, Photo No. 77-5250.

gin, either through the title or content, were used to examine how saw gin makers tried to overcome the limitations of the saw gin and achieve the industry's goal.[8]

Ginners could have been gin makers' most reliable source of information on gin performance, but there is no evidence that the two groups interacted. Both understood the cotton plant and appreciated the risks involved in removing the fiber from the seed. They knew that the fiber was most vulnerable to damage at the moment of removal. That occurred at the first meeting of the organic and the mechanical, when the saw teeth pulled the fiber through the narrow spaces of the grate, wrenching it from the attached seed. Planters maximized successful removal by attending to the preparation of the crop before it reached the gin. It began with planting and extended through the harvest. In an ideal situation, harvesters removed only mature bolls from the plant, then dried it, taking care not to let the seed cotton overdry. Ginners loaded the conditioned seed cotton either directly into the hopper or into the trough of a "feeder" built above the hopper, which regulated supply. Once the gin was powered, the ginner regulated seed cotton supply, adjusted the position of the seed roll, and anticipated and fixed jams. These jobs required not strength but diligence. Closely monitoring the cotton at every processing stage, ginners would have noticed how the saws, grate, and brush interacted with the seed cotton and how they affected nep formation, outturn, and, ultimately, the value of the crop. But whether enslaved African ginners shared their observations with planters or merchant ginners, or whether gin makers would have valued their opinions remains a matter for speculation. Gin makers did listen to planters, some of whom were informed ginners who prided themselves on their familiarity with gin technology.

Ray Collins was such a planter. In 1842 he submitted an article to the *Southern Agriculturist* on his "plan of managing cotton" in order to explain to "fellow-citizens" how he received such high prices. When the price for short-staple cotton averaged 5.7 cents at New Orleans, Collins sold his whole crop, "one bale excepted," for "12½ cents." The excluded bale, his lowest quality, still sold over the annual average at "six cents." Collins let the readers know that the high prices rewarded high quality, which began with attention to planting, cultivating, and harvesting. He cited an article by Mississippi agriculturist Rush Nutt to reinforce the message. Planters next needed to attend to pre-gin preparation such as picking out leaf and twig contaminants and drying the seed cotton carefully. Ginning was the final and most critical stage. He used a common saw gin with "9 inch diameter" saws that had six teeth to the inch. His "grates are straight; and the saw enters them perfectly straight—that is, heel and point at the same time; no false grates; no extra flue."[9] Collins dismissed two popular gin attachments, "false grates," or a set of ribs that sat behind the breastwork or "true grate," and extra flues that either allowed more air to enter the rear of the gin or more fiber to escape it. He concentrated on the single most important interaction in saw gin mechanics, that is, how the saw tooth entered the grate. In emphasizing what is called the ginning point and ignoring brand name or gin size or speed, Collins showed himself to be a planter who was also discerning ginner.

In his dismissal of the popular gin attachments, Collins also presented himself as a conservative, but he reflected saw gin patentees' focus on the grate, the saws, and the ginning point as keys to fiber quality. Patentees studied how the cotton fiber, once seized by the saw teeth, proceeded through the grate. Some believed that fiber breakage occurred when two saws seized the same bundle of fiber at the same time and broke the fibers as they tried to pull the same fibers through different ribs in the grate. When the saws failed either to pull the fibers through or to break them across a rib, a jam ensued, halting ginning entirely. The first problem lowered fiber quality; the second decreased outturn and increased labor costs. Neither was desirable, regardless of cotton prices.

Patentees rallied, designing ribs with ridges and flanges they believed could prevent saws from seizing the same bundle of fiber. Eleazer Carver may have patented the first "improvement in the manner of forming the ribs of saw-gins for ginning cotton" in 1838. He designed a rib equipped with a ridge and a flange. According to him, the ridge raised the surface of the typically flat rib to three effects. It directed the seed roll towards the saws "so that they may with the greater certainty be caught and drawn up by the teeth." Forming an obtuse angle with the side of the rib at the ginning point, it also "prevents the cutting or breaking of the fiber as the cotton is drawn through." The third effect was its flexibility. The design allowed planters to adjust the gin to process the large-bolled, longer-stapled "Mexican" and "Mastodon" cottons.[10] Carver marked the points on the rib where a saw tooth entered the hopper and where it left, then measured the "portion of the circumference of the saw which is between them," and determined that the length of the arc "shall equal the full length of the fiber of the cotton to be ginned." The rib's length should be variable, he suggested, to match variable fiber lengths. He also added a "projection," or flange, on the top of the rib over which the seed cotton slid. Physically separating bundled fibers, it eliminated the breakage problem, according to Carver, and allowed troublesome seeds to pass, solving the jamming problem. Another mechanism on the rib "arrested" the seeds beyond the ginning point but before they were pulled behind the grate, where they would contaminate the fiber.[11]

Carver rewrote the patent and received a reissued patent in 1839 that was half as long and included two new drawings. In the reissued description, he admitted that the saw gin had "been always subject to the inconvenience" of the dual problem of "choking" and fiber breakage, a comment tantamount to an indictment of the toothed principle from one of the saw gin industry's founders. Although the new patent betrayed his doubts about the toothed principle, publicly he remained an advocate of it and he renewed his assertion that a grate equipped with ridged ribs solved the problems. He covered his earlier claims and added construction details and alternatives. The reissue, perhaps intentionally, also included claims that Asa Copeland Junior, a Bates, Hyde & Company gin maker, had already submitted to the patent office.

Copeland worked with his father for Carver's rival in Bridgewater and received the patent for a rib similar to Carver's in 1840. He focused less on fiber breakage than on gin stoppages due to "clogging" or jamming. Fibers and seeds collected in the spaces of the grate, he explained in the specification, and prevented the gin from operating continuously. To "obviate this inconvenience," he designed a projection at the top of the rib that extended above the saw and allowed fibers to pass freely on both sides and seeds to "escape" at the top. Perhaps anticipating an infringement suit from Carver, Copeland emphasized in his formal claim how the rib was to be attached to the hopper rather than its function. When Carver sued, he targeted Joseph A. Hyde, one of the owners of Bates, Hyde, rather than Copeland, although Copeland had not assigned the patent to the company. Hyde's lawyers also stressed the mode of attachment and not the underlying ideas. Carver did not concede defeat when he lost the case in the Circuit Court but went to the United States Supreme Court, where Chief Justice Roger B. Taney upheld the earlier verdict for the defendants in his 1842 decision.[12] Bates, Hyde, and Company celebrated the victory with a broadside featuring Copeland's rib in the foreground.[13]

Undeterred, Carver sued the Braintree Manufacturing Company in 1843 for infringing the same reissued patent. Ironically, Braintree's lawyers subpoenaed John Edson, a mechanic whom Carver had sent to Natchez in 1822, to testify against him. Now an employee of Braintree, he was deemed "interested" in the case and his testimony was ultimately "rejected, as being inadmissible." At one point in the heated trial, Braintree's lawyers accused Carver of having rewritten his 1838 patent expressly to "cover the invention of Copeland," reminding the judge of the recent verdict.[14] The then Circuit Justice Joseph Story ignored the implication that the case before him warranted a similar ruling. "Call the improvement an entirety, or a combination, as we may please," he wrote in 1843, "it is still a patent for 'an improved rib,' and nothing more." Focusing on function, he determined that the defendants had indeed infringed on the reissued patent. He ruled for Carver and denied Braintree's motion for a new trial.[15] Vindicated by Justice Story, Carver had argued for priority, not for the principles he espoused in the patent. What he and the other Bridgewater gin makers really contested, however, was access to a component they believed would solve the quality-versus-quantity dilemma.

Other gin makers offered more complex rib designs. Gin maker John Du Bois from Greensborough, Alabama, designed a one-piece rib that incorporated Copeland's rib projection but extended it into a "back rib" that branched from the flange and entered the region behind the grate. As the fiber passed through the back rib, it was freed from clinging seed fragments or "motes," Du Bois argued. Natchez mechanic C. A. McPhetridge believed the rigid grate was the problem, however individual ribs were configured. In an 1841 patent, he substituted a "grooved roller" that he claimed would prevent "the cutting or otherwise damaging the staple, which it is subjected to by the old arrangement of the permanent grates and revolving saws." His roller would prevent the saws from hitting the grate and throwing off sparks, a known cause of gin fires. In 1858, at a time of high cotton prices, James F. Orr of Orrville, Alabama, introduced a jointed or "sectional" rib that allowed "for the passage of seed and dirt" and the outturn of cleaner fiber.[16] The patent drawing pictured a complicated structure that seemed too bulky to be functional, but Orr put it into production in his Dallas county gin factory, and planters bought it. In the fall of 1857 his brother and agent William K. Orr saw "a large number of [the] sectional ribb gins in operation" and vaunted their success in the pages of the *American Cotton Planter and Soil of the South.*[17]

In saw cylinder and saw design, other gin patentees saw the path to the perfect gin. In 1837 Jacob Idler, a Philadelphia machinist, redesigned the axle on which gin makers loaded the saws. His had a square cross section that prevented the saws from wobbling out of true and striking the sides of the ribs. This, he believed, was the cause of fiber damage. Saws securely mounted and spaced on his axle would "not be liable to cut or hurt the cotton or force it into the teeth, which has heretofore cause the many little knots in the cotton so vexatious to spinners." Furthermore, if the saws "cannot vary they will clean more than is ordinarily done, and deliver the cotton unhurt." The outturn of large quantities of high quality fiber was clearly Idler's goal and he was confident that properly mounted saws could deliver both.

William B. Stewart of Cincinnati believed changes in saw design would achieve the same goal. In 1843 he returned to the notion that saws broke fiber over the ribs. To solve the problem, he arranged variously shaped saw segments in a staggered formation across the axle such that no two saws with the same shape were juxtaposed. Since only one saw seized a bundle of fiber at any one time, they ginned "the cotton with less injury to the staple, and much greater rapidity," Stewart theorized. John Simpson, a carpenter and gin maker in Lewisville, South Carolina, patented a similar idea in 1855. He designed a saw in one piece that alternated toothed with smooth sections "so that no two [saws] can catch the same fibers, and thereby break them." Alexander D. Brown, Israel F. Brown's brother, who may have worked with him at Clemons, Brown, in Columbus, Georgia, patented a saw that alternated smoothed and toothed sections. But he also replaced the circular format with an eccentric shape, achieving in a one-piece saw what Stewart had designed in segments. Brown claimed that the design "rendered [it] impossible for any two saws to catch the same fiber across a rib."[18]

How popular segmented and alternate blank saws were is unknown. Gin makers kept them in production through the period of low cotton prices of the 1840s, beyond the high prices of the 1850s, and into the postbellum economy. The segmented saw was a standard gin component in 1854, according to Mississippi cotton planter and geologist Benjamin L. C. Wailes.[19] In the same year, Carver sold gins "supplied with . . . Saws which are in half segments, and hence easily adjusted," recalling the 1843 Stewart patent.[20] At least a few planters bought Brown's saw through the 1860s. An 1870 Clemons, Brown & Company catalog included his "Alternate Blank Saw Gins," advised for longer-stapled Upland varieties.[21] The fancy saws added complexity at the most crucial point in the ginning process and it is remarkable that planters embraced novelty where they might have retained conventional saws and adopted changes in less risky areas. That even a few patronized the fancy grates and saws is evidence of their confidence in gin makers and of their faith that the changes would yield market rewards.

Gin mechanics held the hopper sacrosanct as they expounded on the problems of fiber quality and quantity in their patents and proffered solutions in the form of redesigned ribs, cylinder axles, and saws. Yet it was in the hopper where seed cotton and gin components interacted. The saws passed through the mass of seed cotton, creating the seed roll by seizing bundles of fibers and pulling them through the grate and off the seed. As ginning progressed, the action of the saws kept the seed roll in motion. But the rotary motion drew the denuded seeds that did not drop out of the hopper to the center of the roll (see fig. 3.2). As ginning progressed, the roll grew in diameter, density, and weight, and it became more difficult for the saws to seize a substantial bundle of fiber and pull it through the grate. The saws only nicked the roll, collecting random pieces of fiber. The ginner intervened with the seed board, a standard hopper insert from the early 1830s, positioning the seed roll at the ginning point, high on the grate, to maximize contact with the saws. It was a random intervention that did not address the underlying problem. Gin makers slowly realized that in order to meet their goal, they had to intervene in the hopper more aggressively.

Two years before cotton prices bottomed, Peter Von Schmidt of the District of Columbia patented a comprehensive redesign of the saw gin that included hopper inserts, a rib modification, and additions behind the grate. He opened his 1846 patent with an explanation of the "difficulties" of ginning and the problems that contributed to low fiber quality. Of all, contamination by "motes, particles of dead leaves, and other foreign matter" most injured the gin and "the staple when passing through the gin with the fibers." Furthermore, the "tenacity with which the fibers adhere to the seed and to each other renders it very difficult to separate them without either cutting or breaking them, or otherwise injuring them by abrasion." Citing the contaminants that harvesters collected along with the seed cotton as a problem for the gin and the fiber, Von Schmidt added that the nature of the fiber itself was also at issue. It clung "tenaciously" not only to the seed but also to other fibers. Wresting fibers free of the seed and of each other resulted in breakage. Moreover, he explained, unless the fibers were "properly loosened" before being "drawn through between the ribs by the saws," they coiled "around the upper part of the rib" and were "there cut by the saws and choke the operation of the machine."[22]

In order to pre-clean and open the seed cotton, Von Schmidt inserted two rotating toothed "cleaners" into the hopper; in order to avoid "coiling," he added a mechanism that vibrated the breastwork, physically shaking the seed roll into place. To further clean the fiber, he included a series of stationary brushes at the bottom of the brush cylinder chamber to "more effectively remove the motes and other impurities." Finally he introduced a "spiral or oblique fan-blower" to increase the air currents that discharged the fiber so that motes might be removed when the fiber left the gin. There is no evidence that this complicated gin went into production, but the patent reeked of crisis.[23] The nature of cotton and the nature of the saw gin were equally culpable in the outturn of low quality fiber, Von Schmidt feared. His remedy was a maze of attachments promised to tame the cotton, revamp the gin, and quell the market.

When North Carolinian gin maker Robert A. L. McCurdy, working in Louisiana, patented his hopper insertion in 1855, the crisis had abated and planters were recovering from the shocks of the 1840s, but the problem of contaminants that Von Schmidt stressed in 1846 had worsened. Some planters directed harvesters to pull the entire boll from the plant rather than to pick out only the seed cotton. Now they needed gins that separated hulls from seed cotton as well as fiber from seed. McCurdy's "Improved Eureka Gin-Stand" accomplished this with a "cylinder screen" insertion that conveyed the hulls out of the hopper before ginning occurred, preventing them from being cut up and contaminating the fiber. The screen was a wire cylinder within which a two-part auger rotated (fig. 6.2). In theory, the seed roll formed around the screen as the ginner loaded the hopper with seed cotton. Since it was attached to the sides of the hopper, it rotated in the same direction as the roll, forcing it into position and keeping it turning. Instead of accumulating at the center of the roll, both seeds and hulls dropped through the screen onto the two-part auger, which "carr[ied] out the hulls . . . thereby effecting a great saving of time and labor and improving the sample."[24]

Hugh H. Fultz of Lexington, Mississippi, a well-known maker of agricultural machinery, patented a similar solution in 1855. Instead of a composite screen and auger, he used "spiral plates," which imparted a spiral and a lateral motion to the seed roll, forcing it to rotate and moving it from side to side. The two motions, he believed, would prevent "the staple or fibers of the cotton from being cut by the saws, and at the same time causing the cotton to be perfectly ginned, or deprived of the seed and motes, and discharged from the machine in separate parts, according to its quality." Acknowledging that the saws "much cut up" the fibers, Fultz designed his plates to discharge different lengths of fiber through different "feed-boxes." Fiber subjected to fewer saw rotations was longer and discharged first at one end, then "medium-sized fiber" was removed from the center, finally "the short at the end opposite." Both seeds and hulls were "stripped" from the fiber and expelled.

A manufacturer and patentee, Fultz put the gin in production and invoked the industry rhetoric by characterizing it as a "perfected" saw gin in a feature article in *Scientific American*. He claimed that gins equipped with spiral plates produced an "increased quantity of cotton ginned by it; second, an improve-

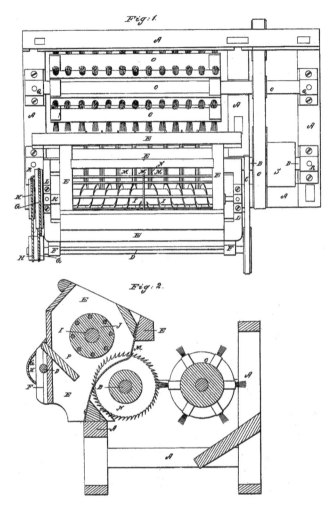

Fig. 6.2. Conservative Change in front of the Grate: Hopper Inserts. R. A. L. McCurdy, Cylinder Screen and Spirally Flanged Shaft, U.S. Pat. No. 13,131 (1855). The top drawing is an overhead view. The two sides of the hopper are labeled "E," between which sits the attachment (I). In both drawings, the brush cylinder is labeled "O." The spiral flange is shown within the cylinder screen. In the bottom drawing, the two attachments (I, J) sit in the hopper (E).

ment in the quality of the cotton over that ginned by the old method—its value being increased from one to two cents per pound, as decided by the cotton brokers of New Orleans; thirdly, all the hulls are discharged with the seed, without being cut with the saws."[25] Producing larger quantities of higher

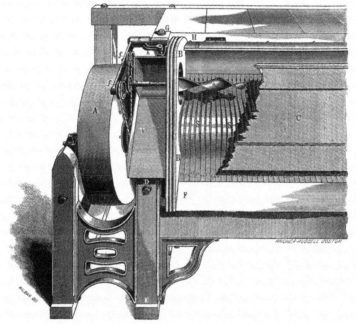

PATENT ROLL-CLEARER,

FOR

COTTON   GINS.

E. CARVER COMPANY,

EAST BRIDGEWATER, MASS.,

AND

CORNER OF ST. CHARLES AND GRAVIER STREETS.

NEW ORLEANS, LA.

Fig. 6.3. Conservative Change in front of the Grate: Hopper Insert. William F. Pratt, U.S. Pat. No. 25,307 (1858). Pratt's roll clearer extended only a few inches into the hopper, the front of which has been made transparent in this trade catalog cover, but it kept the seed roll revolving without McCurdy's heavy apparatus. Courtesy, American Textile History Museum, Lowell, Mass.

quality lint in a strong market, the successful gin demonstrated that rising cotton prices also drove technical change but did not divert the gin industry from its goal.

Hopper inserts remained fashionable through the 1850s as harvesting practices changed and cotton prices rose. William F. Pratt, a gin maker at Eleazer Carver & Company in Bridgewater, patented a hopper intervention in 1859 that revisited McCurdy's cylinder screen idea but simplified it (fig. 6.3). Pratt believed that trashy seed cotton and seed and hull-filled seed rolls were to blame for contaminated, low-quality fiber. He explained that the saw teeth pierced the encumbered roll, cutting the seeds and hulls and carrying the fragments through the grate into the lint room "giving a rough and dirty sample." Solid or hollow cylinders, like McCurdy's, increased the weight of the roll and slowed "the speed of ginning," undermining its purpose. Pratt's remedy was a simple auger inserted twelve inches into the hopper. It was all that was needed to enable the gin "to carry and keep the roll of seed-cotton while in the roll-box in a lighter and more elastic condition . . . and by the same process to remove from the roll the hulls, bolls, trash, and foreign substances without injuring the saw-teeth or cutting or mixing the hulls, &c., with the clean cotton."[26] It could then "loosen up" the dense center of the roll drawing out the heavy seed, hulls, and trash, and discharge them through a "spout."

Like the segmented ribs and saws, Pratt's hopper insertion seemed to add complexity where ginners—and planters—could least afford it. An accompanying two-page brochure published by E. Carver & Company explained that ginners must supply the seed cotton only "at the end opposite of the Roll Clearer, and its operation will be spoiled entirely if the ginner is allowed to feed the Gin in the ordinary manner; and no seed cotton should be fed in, at or near the clearer end."[27] Large gins needed two roll-clearers, one at each end of the hopper. Ginners had to add seed cotton "at or near midway in the roll-box" and "gradually lighter up to the point where it terminates." It warned that "no seed cotton should be fed into the roll-box over that part in which the roll is penetrated by the clearer, and in which it works."[28] Pratt's "roll clearer" was a fragile and fussy attachment that nevertheless appears to have gone into production at a time of escalating cotton prices but changing harvesting practices.

The "huller gin" was another response to cotton collected with hulls and stems attached. Rather than adding inserts to the hopper, huller gin designers added a separate hopper where the hulls were removed from the seed cotton before the seed cotton moved into the primary hopper, where it was ginned. Vicksburg, Mississippi, mechanic David G. Olmstead described its operation in his 1858 patent. The ginner loaded seed cotton with hulls attached into the "huller breast," which was fitted with a McCurdy-type cylinder screen. The saws first rotated through this compartment pulling cotton still attached to the seed through a widely spaced "hulling grate" that removed only the hulls.[29] The saws continued their rotation, pulling the fibers from the seed through a standard grate, ginning the fiber. Gins with huller breasts grew in popularity as the practice of snapping rather than picking cotton spread; they remain in use in the twenty-first century, as ginners cope with machine-picked cotton.

Just as the depression and low cotton prices generated changes in front of the grate—in ribs, saws, and hoppers—they also generated changes behind the grate. In the cavity filled only by the brush cylinder, patentees and gin makers saw an opportunity to treat the fiber after it had been removed from the seed but before it left the gin. The "false grates," which Ray Collins dismissed in 1842 when they were at their peak of popularity, and "mote brushes," were heralded as solutions to poor fiber quality and low prices. A false grate was a set of ribs attached behind the breastwork or "true" grate. The saws loaded with fiber passed first through the breastwork, then through the false grates, which, in theory, removed any clinging seed or leaf fragments. Gin makers installed stationary or rotary "mote brushes" in conjunction with false grates or by themselves; they further cleaned and aligned the fiber before the brush cylinder swept the fiber off the teeth and out of the gin. In 1842, Theodorick J. James of Shirt Tail Bend (now Avon), Mississippi, patented an ambitious combination of these two components that may have contributed to Collins's aversion (fig. 6.4, *top*). He promised that the unwieldy gin would produce fiber "almost entirely free from dust, dirt, and leaf or trash, and consequently every pound will be rendered more valuable than when ginned in the common gin."[30]

Port Gibson, Mississippi, mechanic Hilery H. Kelley avoided some of James's pitfalls by patenting a less complex attachment in 1843 (fig. 6.4, *bottom*). He successfully introduced an attachment that consisted of a series of false grates and stationary brushes that fit compactly behind the standard grate. The saws passed successively through the breastwork and then the false grates before the brush cylinder whisked the fiber from the gin. Kelley facilitated the last process by boxing, or enclosing, the brush and narrowing the vent through which the fiber passed, thereby increasing the draft. He divulged no theory in his patent but introduced the attachment to planters who faced record-low prices. They adopted the attachment immediately and used it to focus attention on global competition, domestic overproduction, and their stake in gin makers' quest for the perfect gin.

In October of 1843, five months after Kelley received the patent award, the *Southern Agriculturist* published a letter from a Mississippi planter, identified only as "A" from Centreville in Warren County, who explained the crisis and its solution. The writer advised fellow planters that they should be alarmed at

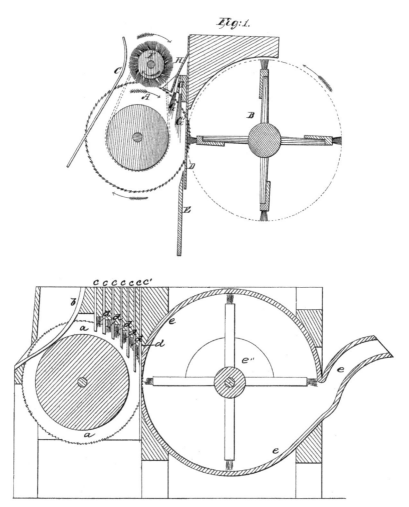

Fig. 6.4. Conservative Change behind the Grate: Mote Brushes. *Top,* Theodorick J. James, U.S. Pat. No. 2,608 (1842). *Bottom,* Hilery H. Kelley, U.S. Pat. No. 3,091 (1843). James and Kelley were two Mississippi mechanics who devised similar solutions to the problem of contaminated lint. Kelley simplified his attachment, facilitating adoption.

the "efforts which are being made by Great Britain to compete with us in the culture of cotton. . . . There is but one course to be pursued," he wrote. Planters must "excel them in quantity and quality." If they were to continue to "send forward large quantities of inferior cotton," they would soon lose market dominance. "Quality" depended "upon the manner in which our crops are gathered, but," he continued, "we are satisfied that the perfection to which mechanics are rapidly tending, will have a vast influence on the character of our cotton."[31]

"A" cited Kelley's attachment as an example of "the perfection to which mechanics are rapidly tending." The writer had accompanied "several neighbors" to the plantation of a judge who had installed the attachment on his gin. He described it as consisting "of four sets of false grates, two stationary brushes, and a 'boxing up' of the main brush." The judge arranged a trial, which "A" reported "was highly gratifying to all present. . . . Not a moat or false seed was found after having ginned 60 or 70 lbs. seed cotton." Gratified by his investment, the judge believed that the attachment added "from a half to a cent on the pound" to the value of his cotton. "The improvement has been patented," put into production, and "cost $75."[32] It was available and affordable, and, "A" intimated, should be adopted by sensible planters.

In the patent drawing, Kelley had illustrated seven sets of false grates and six stationary brushes, but had introduced a four-and-two combination that "A" saw on the judge's gin. He continued to simplify the attachment; in December, the *Southern Agriculturist* reprinted an article from the Tuscaloosa, Alabama, *Monitor* that described it in a three-and-two configuration "through which the saws pass in their revolution as closely as possible without friction." Discerning planters rejected his boxed brush and narrow vent, recommending that a "'grated flue' . . . through which the dust that may pass the brush wheel may fall in its passage to the [lint] room" be added instead.[33] Within seven months of receiving the patent, Kelley had introduced the attachment to Mississippi and Alabama cotton planters. His success lay not only in his mediation of the crisis planters faced and in their eagerness to adopt novelty but also in his willingness to adapt to planters' competencies.

Bridgewater gin makers simplified the "false grates" into an attachment they called a "guard" and installed it on their popular gins. In 1844 Eleazer Carver patented a device "designed to protect and hold the fibers of cotton while on the teeth of the saws . . . to enable the brush to operate more perfectly in taking the cotton off from the teeth of the saws in loose and uniform quantities." In the standard gin, Carver explained, the brush removed fiber that was "bent backward around the teeth of the saw" instead of separating and straightening it. The wind generated by the brush exacerbated the problem by removing clumps of "doubled, connected" fiber in an "improper state." His "guard" was a simple flap of metal that hung between the saw and brush cylinders. It cut down on the draft and allowed the fibers to remain on the teeth long enough for the brush to "perform its proper operation," detaching small amounts of fiber and straightening them in the process. Used in combination with the boxed brush cylinder that he patented in 1845, the guard became a standard component of the Carver Gin.[34]

In 1849 Carver agents in Mississippi, Tennessee, Louisiana, and Alabama announced the availability of "a very superior stock of gins" loaded with the new attachments. Carver gins, they informed planters, now included "a very great improvement in the Cast Iron back Grates (for moting Breasts) by which nearly all the motes and trash is taken from the Cotton, with comparatively no loss [of fiber]. The[y] also have a large Drum Brush, with thirty-six rows of Brushes by the construction of which the cotton is taken from each tooth separately and presents the appearance of having been carded." Translating Carver's ideal into reality, the improvements promised to increase the quality of fiber without sacrificing quantity. Agents added an incentive to skeptical planters, offering to install and adjust the gin and give it their "personal attention . . . free of charge."[35]

The reception of the already-popular Carver Gin with the attachments prompted rival Bates, Hyde & Company to add a guard and mote-brush attachment to its Eagle Gin in 1846. The patentee was gin maker Edwin Keith, who incorporated Asa Copeland's rib and only slightly modified Carver's drawing and description. Baiting Carver by flaunting the contested rib, infringing his design, and simply paraphrasing his patent, Keith failed to goad him into another expensive lawsuit.[36]

A regional style emerged within this class of modifications that spanned periods of depression and prosperity. Bridgewater gin makers favored compact and simple attachments whereas southern gin makers introduced elaborate false grates and used them in combination with stationary and revolving mote brushes. Mississippi gin maker Theodorick James pioneered in 1842 with his complex gin. Alabama gin maker John Du Bois installed a rotary mote brush cylinder over the split rib in his 1850 patent. Arkansas gin maker Thomas J. Laws added "wings" to his small mote brush in an 1852 patent; according to Laws, these "obviate the great defect of the common cylindrical mote-brush which does not throw off the motes separated from the cotton and consequently soon clogs up and becomes useless." By adding revolving cylinders rather than stationary brushes behind the grate, southern gin makers introduced what evolved into a craze in the late 1850s.

Through the decade, gin makers added more and larger rotary mote cylinders, building large multi-axle gins that turned out greater quantities of fiber. Like the hopper insertions, these changes emerged out of the panic of 1837, but gin makers elaborated them in the bullish economic environment of the late 1850s. Technological momentum drove them to build increasingly complex gins as they perfected their machine skills. The market also impelled change as rising prices attracted new cotton planters and pushed others to expand production. Cotton is not a time-sensitive crop; it can be harvested, ginned, and stored in bales. However, marketing cotton was time-sensitive—planters who sold in global, not only national markets, needed to have bales ready to ship when prices were favorable. As more planters became concerned with processing ever-increasing amounts of seed cotton within time constraints, they made gin speed and outturn overriding concerns. Planters addressed speed on one level by converting from mule and water mills to steam engines to power their gins. On another, they urged the construction of larger gins that they could run faster. The push for quantity forced a redefinition of quality. Over the decade of the 1850s, fiber cleanliness, not length, became a critical marker—but outturn trumped both characteristics.

The multi-axle gins that became the rage among Mississippi Delta planters in the late 1850s were responses to the redefinition of fiber quality and to the demand for larger, faster gins. The standard saw gin included two axles, or cylinders; one carried the saws and the other the brush. A multi-axle gin included at least one more cylinder, a behind-the-grate rotary mote brush. These complex gins signified technological enthusiasm as much they did a practical response to the booming short-staple cotton market. But culture rather than crop size may better explain why Mississippi planters were attracted to multi-axle gins and why Mississippi gin makers patented and made the majority of them. Out of nine identified patentees or manufacturers, Alabamians John Du Bois and Daniel Pratt were two of only three exceptions. Du Bois patented and made his gins in Greensborough, Alabama, and Mississippian David G. Olmstead contracted Pratt to make his ambitious "Eureka" gin in Prattville. Alabama-born gin maker Thomas J. Laws, who patented out of Hempstead, Arkansas, was the third exception. One multi-axle gin patentee who may not have put his gin into production was J. Alexander Ventress, in Woodville, Mississippi. Complexity appealed to Mississippi gin makers and planters alike; in it they saw the promise of perfection.

Of the multi-axle gin patentees and makers, Benjamin David Gullett is the best known. In the 1830s he had left his home in North Carolina and traveled to the Deep South. Like Samuel Griswold, Daniel Pratt, and T. G. Atwood, he worked first as a carpenter before he began making gins. He did not tarry in Georgia, but settled in Eutaw, in Greene County, Alabama, in the late 1830s. He met and married his wife there and was known for designing and building an impressive Baptist Church in 1843.[37] He may have met gin makers Charles Johnson and William Avery, who advertised out of Erie in 1830, but he certainly knew Samuel R. Murphy, a gin maker and planter who lived in Eutaw. Gullett could have learned the art of gin making from him, or he could have visited the nearby factories of John Du Bois and Pinckney Jones in Greensborough.

Opportunity pulled Gullett west to Aberdeen, Mississippi, in the mid- to late 1840s; he may have had family there.[38] Whether he joined William T. McCracken at his Aberdeen Gin Factory & Machine Shop or opened his own small shop, 1849 saw Gullett launch his gin-making career.[39] He gave no occupational title to census marshals in 1850, but they recorded the five enslaved Africans he owned, one of whom was a working-aged man, and real estate valued at $35,000. One source reported that he worked with wealthy merchant William R. Cunningham and gunsmith Robert Kirkpatrick, who bought out McCracken in 1851, possibly in the wake of his labor unrest, and that it was they who operated the Planters' Machine Works.[40] Although Gullett appeared to have restricted his interests to gin and press design, affiliation with the firm would have given him access to a foundry and an array of modern machinery, woodworking equipment, and tools.

Tennessee-born Leonard Campbell worked in a similar setting in Columbus, the seat of adjacent Lowndes County, and invented a similar gin. Five years younger than Gullett, he was also less wealthy. He owned one young enslaved African woman but claimed no wealth in 1850. R. G. & Dun agents logged him as without "means" and not "respons[ible]" in 1854. They thought more of his partner, the founder Thomas W. Brown, who may have been the same person who formed a partnership with Cyrus S. Aikin in 1840.[41] The agents described him as a "money lender, & note shaver" worth "not less than $20m clear of all indebtedness." The partners had opened "a new firm, formed to carry on Manuf[acturing] of a new kind of Cott[on] Gin invented by Campbell which is tho[ught] to be superior to any wh[ich] has preceeded it. It has been lately patented."[42]

Both Gullett and Campbell received patents in 1854 for gins with be-hind-the-grate attachments. Campbell combined a stationary mote brush and guard with an attachment he called a "stationary concave." A pre-patent award summary published in *Scientific American* described the "improvement" as a "series of passages through which the ginning saws work." Campbell made them out of a grooved piece of solid metal, lined the grooves with bris-tles, and attached it behind the breastwork. As the saws carried fiber through the "passages," the bristles freed "the cotton from impurities." Like the Carver and Keith's guards, it also reduced the draft generated by the brush cylinder, allowing the brush time to dress the fiber. He found the design wanting and in 1855 took another patent that added a large rotary mote brush, which he lo-cated beneath and tangent to both the saw and brush cylinders. The combina-tion, he wrote in the specification, "perfectly ginned and cleansed" the fiber before it was discharged from the gin. The partnership of Campbell & Brown, to which the patent was assigned, had tested the cylinder. Campbell declared that it "renders perfect the gin formerly patented by me."[43]

On his own search for perfection, Gullett began more aggressively than Campbell. In his 1854 patent, he combined a stationary and a rotary mote brush (fig. 6.5, *top left*). Since both components had been previously patented, Gullett claimed priority only for the combination. Since fiber cleanliness was more an industry concern than length at the time, he emphasized that his brushes "separated the motes from the cotton before the latter leaves the saw," cleaning the fiber "and otherwise dressing it better." In 1856 Gullett formed a partnership with Richard Gladney styled Gullett, Gladney & Co.[44] Encouraged by the reception the gin received, Gullett took the brash step of patenting and introducing a four-cylinder "Steel-Brush Saw Gin" in 1858 (fig. 6.5, *top right*). He placed the "carding cylinder" beneath the brush cylinder and the "steel comb brush" behind both. The last largest brush cylinder re-moved the fiber from the two smaller cylinders and blew it from the gin, aided by "blast boards" built into the sides of the frame.

Gullett's behemoth thrilled planters who relished novelty and complexity. Noah Cloud, editor of the Montgomery, Alabama–based *American Cotton Planter and Soil of the South* inserted a notice in an 1858 issue extolling the "Patent Steel Comb Brush Gins." The gins "have a high reputation, and are giving entire satisfaction to those using them," adding "one-half to one cent" to each pound of fiber. Hoping to interest Alabama planters, he added that "Gullett, Gladney, & Co." would install a stand at the company's headquar-

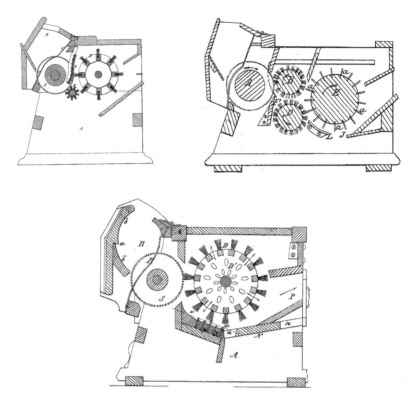

Fig. 6.5. The Evolution of the Gullett Gin. Benjamin D. Gullett, U.S. Pat. No. 10,406 (1854), No. 19,417 (1858), No. 140,365 (1873). Gullet added stationary mote brushes (H) and a rotary moting cylinder (F) in his first patent shown on the left. Four years later, he introduced a gin with four cylinders, on the right. It featured conventional saw (A) and brush (E) cylinders, his "carding brush" (B), and the "steel brush" (D). Fifteen years later, he retreated from complexity with his 1873 gin, equipped only with "beater blades" (g) below and to the left of the brush cylinder (D).

ters for them to inspect.[45] Martin W. Philips, planter and the owner of the Southern Agricultural Implement Company in Hinds County, Mississippi, contributed frequently to southern agricultural journals and freely shared his opinions on many topics, among them cotton gins. Before Gullett introduced the "Steel Brush," he had extolled Bridgewater gins, although he patronized Mississippi gin makers Jacob Hewes and T. G. Atwood. Gins made anywhere between Massachusetts and Mississippi, Philips deemed unworthy of his consideration. After complaining about the high cost of the Gullett gin in an 1860 article for the *Southern Cultivator,* he announced that he had bought two, adding praise for the quality and quantity of fiber they turned out. "The Gullett Stand has two brushes and a steel comb on a cylinder [in addition to the saw cylinder], yet I have had no more trouble than with other Stands."[46]

Louisiana planter G. D. Harmon testified that "every planter without a single exception, who has witnessed the performance of the Gullett Stand, ties the *blue ribbon* upon it. They can not only gin *faster* than any other stand, but they make a *better sample*." Powered by a steam engine built by "E. Barbaroux, of Louisville, Ky., and put in motion by that energetic, intelligent machinist, Thomas Stout," Harmon's two eighty-saw Gullett gins each ginned twelve four-hundred-pound bales of cotton per day, four more than his Bates, Hyde "Eagle" stands.[47] Harmon celebrated southern ingenuity as he pitted his southern-made machinery and mechanics against the once-coveted northern gins. As sectional tensions rose, promoters dubbed Gullett's gin "The Great Southern Gin" and touted it as unrivaled "for speed, lightness of draft, and beauty of sample."[48] It would be the pinnacle of complexity for Gullett, who retreated to Zen-like simplicity in an 1873 patent for a conventional gin with minimalist grid bars at the bottom of the brush chamber (fig. 6.5, *bottom*).[49]

Gullett never ventured in front of the grate but other gin makers who adopted multiple axles added huller breasts and fancy grates, building gins that effused technical proficiency. A. Q. Withers of Byhalia, Mississippi, designed a curved roll box with a springboard that compressed when the seed roll "increases in size" and expanded when it decreased. "By this arrangement," he explained in his 1858 patent, "the seeds are retained in the roll-box till they are completely stripped of 'lint'; while at the same time the stripped seeds, hulls, and trash are allowed to escape." He also "interposed" behind the grate a second but smaller set of saw and brush cylinders, and used guards to cut the air velocity. Stationary mote brushes further dressed the lint and a rudimentary condenser removed any remaining motes and compressed the lint into a batting. "With these improvements," he concluded, "the cotton comes from the gin in a greatly improved state, being almost entirely freed from impurities, and resembling carded cotton rather than the matted bunches frequently coming from ordinary gins."[50]

Perhaps the most ambitious of the gins that were put into production were designed by David G. Olmstead of Vicksburg and built by Daniel Pratt in Alabama. In 1858 Olmsted had patented a multicylinder gin with a huller breast but the one he patented in 1859 was more complex. Behind the grate, he added stationary brushes and "extensions" that cut down the draft generated by the brush. In front of the grate, he moved the cylinder screen from the huller to the plain breast and added a "hulling grate" through which dehulled seed cotton passed on its way to the breastwork. He also incorporated flanged ribs to direct the seed cotton to the ginning point. Priced at six dollars per saw, compared to the three dollars per saw that Campbell & Brown charged or the five dollars per saw that Gullett charged, Olmstead's three-cylinder, huller gins were expensive as well as complicated. Still, "Delta planters" coveted them and Pratt scarcely met demand.[51]

By the end of the 1850s, Delta planters signaled another shift in the definition of fiber quality, which the complex gins embodied. Where once it had signified staple length and cleanliness, quality had slowly became synonymous with quantity. Martin W. Philips charted the change. Reflecting on opinions he had published in the late 1840s and 1850s, Philips wrote in 1860, "Ten years ago, I believed 1½ bales and a No. 1 sample went hand in hand. I have seen my error."[52] He once thought, he explained, that low outturn was the necessary sacrifice for high quality. Gullett's multicylinder gins had changed his opinion. In the 1840s he had likely used forty-saw gins; by the end of the 1850s, the average size had doubled. In 1860 Philips ran a mule-powered sixty-five–saw Gullett gin, equipped with an automatic feeder made by Jedediah Prescott of Memphis, Tennessee. With the help of a "small woman," the gin turned out "5 to 6 [400-pound] bales per day," he claimed. Planters still concerned with fiber length charged the gin with "cutting the fiber," but Philips ignored them. The gin delivered the quantity he wanted, and he declared the "lint equal to any that goes into market."[53] Harmon echoed Philips in his focus on merchantable bales and lint appearance rather than on the minutiae of fiber characteristics that had concerned gin makers and planters in the 1840s.

Gin makers, even those convinced that fiber quality was the key to high prices, shared planters' interest in outturn and had increased the capacity of the saw gin. The changes that led to greater quantity generated modest innovations in axle and bearing construction, lubricant choice and dispersal, as well as frame design, but the effects were measurable. In 1796 Phineas Miller averaged one hundred pounds of fiber per day at his animal-powered Mulberry Grove ginnery, one fourth of a mid–nineteenth century bale. Gin makers had doubled that amount by 1820, William Aikin claiming in 1821 that his gins turned out approximately two hundred pounds, or half a bale. In 1834 Sturdevant & Hill, a gin factory in Selma, Alabama, warranted its fifty-saw gins capable of turning out thirteen hundred pounds of fiber, or three and one fourth bales. Thomas W. Brown promised in 1856 that Leonard Campbell's gin of the same size could gin five to six bales. Powering his two eighty-saw gins with a steam engine, G. D. Harmon exceeded gin makers' and planters' most sanguine expectations.

Through bear and bull markets, these men remained committed to the saw gin and to the idea that they could perfect it by altering and adding components. They succeeded by redefining "quality"—and, with it, "perfection." At the beginning of the period, they defined quality fiber as fiber that emerged unbroken through the ginning process. The perfect gin was equipped with flanged ribs, segmented saws, and boxed brushes to assure its safe passage. At the end of the period, they defined quality fiber as clean. Cleanliness had always been an industry focus and had concerned J. R. Bedford when he described an "invaluable improvement" that eliminated "trash and dirt from cotton" in 1821. But it became an overriding issue in the late antebellum period. The shift in the definition of quality and in the conceptualization of the perfect gin within the saw gin model was an admission of the inherent inability of the saw gin to preserve fiber length. Gin makers conceded defeat and focused on a problem they could solve. The saw gin of 1860 delivered clean fiber in large quantity and saw gin makers celebrated the attainment of perfection.

CHAPTER SEVEN

Old and New Roller Gins, 1820–1870

The low cotton prices that accelerated incremental change in the saw gin prompted a reexamination of the roller gin. One group of gin makers, located primarily in the North, introduced modernized roller gins for long-staple cotton. Another group with northern roots applied roller gin principles to two radically new gins for short-staple cotton, the McCarthy gin and the cylinder gin. Both groups pursued the industry goal of perfection in machines designed to maximize quality and quantity. Modernized roller gins turned out more fiber but they inevitably sacrificed quality, a fatal defect in a market that privileged quality. The McCarthy and cylinder gins turned out longer, cleaner fiber but failed in their intended market because both gins sacrificed quantity in a market that privileged quantity. Despite the modernized roller gins, the foot gin persisted in the long-staple cotton market while the McCarthy and cylinder gins never threatened the saw gin in the short-staple cotton market. Roller gin makers and long-staple cotton growers claimed success in the quest for perfection after they modified their definition of quality and declared that the McCarthy gin was the machine that delivered it.

The bifurcation of the cotton industry in the 1820s into saw gin users and roller gin users reinforced competition between the associated crop cultures. Planter-politician and roller gin advocate Thomas Spalding heightened it with his portrayals of short-staple cotton growers and saw gin users. Writing for the *Southern Agriculturist* in 1835, he characterized roller gin–using, long-staple Sea Island cotton growers as men like himself. They were "generally an educated people, and a stationary one, less anxious after change than their countrymen are supposed to be, and although severely smitten . . . in peace by the National Tariff, they have still clung with some degree of fondness, to the places, whereat they were born, and to the seas, in which they were bred."[1] The short-staple cotton producer was, in contrast, devoid of place. "Wherever he roams, however unfortunate his condition, the cotton seed, like the maze seed, is carried along with him," Spalding pronounced, intending to disparage the white planters by likening them to nomadic Native Americans.[2]

"Why then," Spalding asked, "has a whole people so readily and so greedily adopted a course of cultivation leading to so small individual benefit?" "There are many reasons," he answered, why newcomers grew short-staple cotton, "but there is one, sufficient to the end. It suits their wandering habits, it requires no great time to grow, it requires no great expense to prepare, and it will command money, less or more, wheresoever grown." The saw gin suited this type. Invented "to do much in a short time [it] was strong enough to travel without being broken to pieces, and light enough to move with its moving master." Derived from the "cylindrical whipper, and the circular cards," he believed, "Miller and Whitney's gin . . . diminishes the quantity as well as injures the quality" of the cotton, leaving more cotton on the seed than if it had been roller ginned, and "cutting" it as well. Still, Spalding conceded, it represented "a cheap and expeditious mode of taking the wool from the hairy American cotton."[3]

Nostalgia informed Spalding's contempt for short-staple cotton growers who threatened coastal planters' hegemony, but it distorted his representation of long-staple cotton growers. They responded to the panics of 1819 and 1837 with "expeditious" changes in planting and processing techniques aimed at lowering costs and increasing profits, just like the short-staple cotton growers he maligned. Long-staple cotton growers received more for their prized fiber than did short-staple cotton growers, but they also experienced steeper drops. For example, in 1819 Upland cotton sold in Liverpool for a low price of 10 pence per pound whereas Sea Island received 21 pence. In 1828, the year long-staple cotton growers mobilized to stem the decline, the fiber dropped eleven pence but Upland cotton only five. Prices rebounded until the credit markets collapsed in 1837. That year Upland cotton sold in Liverpool for 7 pence per pound to Sea Island cotton's 16 pence. By 1840, when the McCarthy and cylinder gins were introduced, Upland cotton had fallen 1.25 pence but Sea Island 2.50 pence from 1837 prices.[4]

When South Carolina governor and long-staple cotton planter White-marsh B. Seabrook published his *Memoir on the Origin, Cultivation and Uses of Cotton* in 1844, prices in both markets had continued to decline but Sea Island cotton producers like him lost more than their short-staple cotton counter-parts. Upland cotton sold for 2.37 pence less than it had in 1840 but Sea Island cotton had dropped 3.3 pence. The declines shook long-staple cotton growers. They responded like their short-staple colleagues, with calls for increased at-tention to husbandry, processing, and packing, and even for voluntary acre-age reductions.[5]

The *Southern Agriculturist,* founded in Charleston by John D. Legare in 1828, began as the mouthpiece of the long-staple Sea Island cotton commu-nity. In the first issue, "A City Rustic" blamed overproduction, and an anony-mous writer, "illiterate managers," for the "unexampled depression" of cotton prices over the past decade. "Everything now depends upon the excellence of the article, and the skill employed in its preparation. . . . Men of leisure and in-telligence" must replace haphazard practices with "accurate observations and well-conducted experiments," the author lectured, calling for empirical re-search to replace tradition.[6] Other writers provided guidelines. They typically began their articles with practical tips on soil preparation, seed selection, cul-tivation, and harvesting. After discussing how best to dry and pre-clean seed cotton, they inevitably turned to the gin.

Long-staple cotton growers were committed to the roller gin, the only tech-nology that removed the full length of the fiber by pinching it from the seed, but they were exasperated by its limitations. Two types had survived the tran-sition period. Most merchants and planters with large crops had replaced the labor-intensive barrel gins with self-feeding Eve gins, but some former Eve gin advocates returned to the foot gin.[7] Spalding, who had once distributed the Eve gin, now disparaged it, joining a chorus of critics. "Much money had been spent upon costly machines, propelled by horses, by water or by wind, first in the Bahama Islands, and for many years in Georgia and Carolina," he wrote in

1835. "But at last most of the growers of sea island cotton have returned to their first and most simple machine, to wit, two wooden rollers kept together by a wooden frame" powered by foot.[8] Planters preferred it because it preserved the staple in a market that rewarded staple length. But at times of low prices, they criticized it for its low outturn.

Gin makers believed that if they could solve the four major problems associated with the roller gin—lapping, springing, supply, and power—they could build a gin that preserved the staple but turn out more of it. Lapping occurred when the fiber, after being pinched from the seed, stuck to the rollers and revolved around them instead of dropping off. Multiple passages scorched the fiber, discoloring and weakening it while also clogging the gin and sparking fires. Springing defeated ginning altogether. It happened when the rollers, necessarily narrow in order to pull off the fibers without crushing the seed, sprung apart at the center when the ginner loaded the gin. Joseph Eve had addressed supply in 1788 with his self-feeding gin, but the feeder technology did not transfer to other roller gin types. Supply had remained a processing bottleneck—literally in the hands of the ginner; the power source had remained in his feet. Gin makers and planters had used animal-, water-, and even steam-powered roller gins and were confident that if they solved lapping and springing problems, they could automate supply, and power their plantation-made foot gins with a more consistent, reliable source.

Gin makers responded with changes that complicated a machine valued for its simplicity. The earliest was a patent held by a British citizen who addressed the problem of lapping. James Harvie identified himself as an "Engineer, late of Berbice," a contested British colony in present-day Guyana on the northeast coast of South America. In 1820 he patented a method of doffing the fiber from the rear of roller gins by "shifting brushes," thereby avoiding reliance on gravity to shed the fiber. He published a summary in an 1824 issue of a London journal of "specifications of patent inventions" in which he described how to build and install the brushes. They were to move laterally when the rollers turned, preventing the fiber from "being carried round the rollers, whereby it has hitherto been subjected to great injury in its colour and fabric."[9] But the brushes heated in the process, Seabrook noted in his 1844 *Memoir on Cotton,* suggesting that the invention had gone into production but failed. He also mentioned a United States patent for hollowing out the rollers and passing "cool air, or even water" through them to eliminate the "liability."[10]

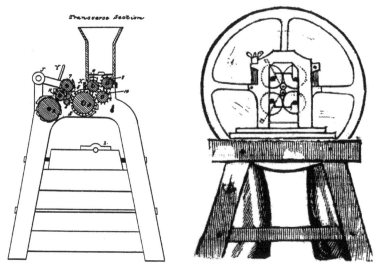

Fig. 7.1. Modernized Roller Gins. *Left:* Jesse Reed, U.S. Pat. Unnumbered (1827). Reed used four spiked feeding cylinders (right) and three doffing rollers (left) in his cast-iron gin frame. The ginning rollers are the two smallest circles in the drawing. *Right,* William Whittemore and William Whittemore Jr., U.S. Pat. Unnumbered (1834). The Whittemores also relied on manual supply but introduced friction rollers shown as four large circles that surround the two small ginning rollers. The halo around the structure is the drive wheel that could be attached to a treadle or a mill. *Southern Agriculturist* 8 (1835): 474.

Jesse Reed, a machinist from Bridgewater, Massachusetts, introduced a complicated solution in his "Sea Island Cotton Cleaner," which he promised would eliminate springing and automate supply (fig. 7.1, *left*). Its "prominent characteristics," he wrote in the 1827 patent, were "its simplicity, efficiency, capacity of feeding itself without the danger of clogging and of discharging cotton with certainty when separated from the seeds." To that end, Reed designed a cast-iron frame that held nine cylinders, each built with metal cores to prevent warping or springing. A narrow hopper sat over four spiked feeder cylinders, which "received, separated, and conveyed" seed cotton to the ginning rollers. He made the uppermost roller out of fluted or ridged steel and wrapped the lower one "obliquely with a strip of roughened leather" to form a "firm smooth surface." Like saw gin makers, he built a large brush cylinder geared to rotate faster than the ginning rollers that doffed or removed the fiber. Two spiked cylinders, placed between the ginning rollers and the doffing cylinder, prevented lapping. Reed had learned valuable lessons from an ill-fated 1826 patent, he wrote. He felt he could now claim "the difficulty of cleaning Sea Island Cotton with dispatch by machinery without injury to the staple was never surmounted until the time of [this] improvement." Reed introduced the gin to South Carolina planters, who praised it for its "superior workmanship" but deemed it "too complicated and expensive" for their purposes.[11]

William Whittemore and his son, William Jr., of Cambridge, Massachusetts, patented a roller gin promised to eliminate lapping, springing, and overheating. It too had limited success, but it generated interest among articulate planters who used it to emphasize the challenges they faced. The Whittemores were members of a prestigious family of machinists. Whittemore Sr. was a younger brother of Amos Whittemore, the celebrated inventor of a machine that automated card-clothing manufacture. He developed the famous 1797 patent at the firm William Whittemore & Company, where the brothers made card cloth for machine and hand cards. When the company reorganized in 1818, William's brother Samuel bought his share and William left to pursue other interests.[12] The association with card-clothing manufacture did not lead him to the saw gin, as might be expected. Instead, in 1833, he and his son copatented a roller gin.

Like Reed, the Whittemores adapted textile mill machinery to solve the problems of the roller gin. They ran "endless aprons," similar to conveyor belts, around the ginning rollers to power them and possibly also deliver seed cotton. A sheet of metal with a tapered edge called the "guard" was installed tangent to the upper roller to doff the fiber. It solved the lapping problem without adding a complicated mechanism like Harvie's 1820 "shifting brushes." Thumbscrews mounted at the top of the frame, reminiscent of Mezy Le Normant's 1746 "Moulin de Prat," regulated roller pressure and allowed the ginner to make fine adjustments. Whittemore Jr. revised it in subsequent patents taken in 1834, 1835, and 1839.

"Whittemore's Improved Cotton Gin" caught the attention of planters and the southern press before its final 1839 design. In May 1835 an unidentified planter declared it "the greatest improvement on the Sea-Island Cotton Gin" and highlighted its features in a brief article reprinted by the *Southern Agriculturist*.[13] In September the journal reprinted a longer article, this one "from the *Charleston Courier* of June 13, 1835," that added empirical evidence to its endorsement of the Whittemore gin. "This is emphatically an age of enterprise and improvement, as well as of invention," a planter identified only as "J. B. W." began. Sea Island cotton planters had long desired a gin that required "less expenditure of time and strength than is ordinarily required by the common old fashioned roller foot gins." Faced with low prices in an increasingly competitive market, they now demanded one. Thirty pounds a day was the "task for a negro" on a gin with iron "balance wheels," twenty-five with "wooden balance wheels," sometimes less, he explained. After harvest, "the strongest and best hands upon a cotton plantation," who might be more profitably employed in the fields, are "occupied for about three months in the year, in ginning out the crop of the last season." Eve's gin was not used on the finer staples now grown, nor was Reed's gin suitable—but the Whittemore gin had potential.

Whittemore had "three times visited the South in reference to the wants of this community" and had built several models based on planters' suggestions, after which he had "altered, amended and improved" the gin, J. B. W. informed readers. The inventor now believed that he had "produced a machine for ginning fine sea-island cotton, superior to any thing of the kind ever before invented, and one capable of no further improvement." J. B. W. conceded that it was the "most simple in its construction, and on that account, extremely well adapted to the use of negroes." Equipped with rollers "nine inches in length and one and one-sixteenth in thickness, it was a durable machine that ran smoothly with friction rollers. These prevented the cast steel and hickory rollers from heating; "guards" prevented lapping (fig. 7.1, *right*). J. B. W. disliked Whittemore's treadmill arrangement for animal power, preferring to use the "old horizontal cog wheel and trundle," and he questioned the wisdom of restricting the ginner in the foot-powered model to the use of only the right foot. He preferred to attach the crank to allow the ginner to use either or both feet. But these were minor criticisms. "As to the *style* and *manner* in which Mr. Whittemore's gins perform the operation, there is no doubt, there can be no doubt, but every honest judge, and every candid cotton buyer, will readily admit, that they deliver the cotton uninjured in its staple."[14]

The effusive endorsement echoed those of other boosters, but then J. B. W. hedged. The gin ran fast, but he wondered if speed impaired fiber quality. In an animal-powered model, the rollers revolved "265 times a minute" but could attain "285" if the horse was "urged to a rapid step." He questioned if "more than 250 revolutions in a minute are consistent with good ginning." Over that number, seeds were apt to be crushed and the oil stain the fiber. Furthermore, "where there are no friction wheels the machinery becomes heated," suggesting that there were circumstances where friction rollers could not be used. Moting lint that was peppered with seed fragments and discolored fibers was time-consuming and expensive, increasing costs and lowering profits.

Planters had struggled in vain to improve outturn, J. B. W. explained. They experimented with larger, longer rollers that allowed more seed cotton to be processed, but the thicker rollers crushed the seeds and the longer, narrower rollers sprang. Experiments with different roller materials had only partially succeeded. Steel rollers, once touted by prominent planters like Kinsey Burden, could be made long but were "liable to cut the cotton, and of course injure its staple." From these disheartening facts he deduced that part of the solution to the problem of outturn lay in supply. J. B. W. acknowledged that "Col. Reed" had made a failed attempt "somewhat upon the principle of Evans' gin," meaning Joseph Eve's gin, but urged improvement: "Some means must be devised for feeding the rollers, by the substitution of machinery for human hands." In the absence of alternatives, he declared "Mr. Whittemore's gin . . . the most complete and perfect piece of machinery which has ever been contrived for the specific purpose of separating the seed from Long Staple Cotton."[15]

The testimonial seemed to signal continued success for the gin. *Southern Agriculturist* editor J. D. Legare added a note to J. B. W.'s article alerting planters that "Mr. Whittemore" had sold machines "from Carolina to Florida" and had returned "to the North" to procure more and expected to have them for planters "in time for the crop of the present season." Yet nothing more appeared about the gin in the journal. In his 1844 *Memoir on Cotton,* Seabrook explained the silence. Planters had found that the gin did in fact "cut the staple." In a market where fiber length remained synonymous with quality, planters considered it an "irremedial defect," and the gin was "consequently abandoned."[16]

Where the William Whittemores attempted but failed to overcome the limitations of the roller gin—particularly supply—Joseph Eve succeeded. His large self-feeding roller gin had its critics, but merchant ginners and planters with large estates used it, now only on coarser long-staple cotton crops (see fig. 2.3). Legare may have been responsible for a revival after he published a two-part account of his "Agricultural Excursion made into the South of Georgia in the winter of 1832." He called on plantations of prominent long-staple cotton planters, visiting Thomas Spalding at his Sapelo plantation, John Couper at Hopeton, and James Hamilton at Hamilton. Noting the details of planting and cultivation, he studied the processing stage closely. Enslaved African ginners used Eve gins without difficulty, he noted. With a duplex gin, one with two sets of thirty-inch rollers, one ginner turned out "from four to five hundred weight per day" at Hamilton. "Somewhat at a loss" to understand why so efficient a machine had been "tried and abandoned" in South Carolina, he suggested that planters reintroduce them.[17]

James Hamilton Couper, John Couper's son, managed James Hamilton's plantation and gave Legare additional information about the Eve gins he and area planters used. They built gin houses about "thirty-four to thirty-six feet square, two stories high" to hold two gins. They put the power source, then a "horse walk," on the ground floor, and the gins and screened moting tables above it on the first floor. The whipper and "a shop for turning rollers" went into the garret.[18] John Couper and Major Pierce Butler, another neighbor, had installed Eve's duplex models in their gin houses. Legare considered them "somewhat complicated" but marveled at the feeder.[19] "As there is a constant motion upwards and downwards, the cotton is generally in such a state, as not to choke the rollers." He noted that it raked open the compact bolls before presenting them to the rollers. He saw enslaved Africans turn replacement rollers on the lathes but they did not make the gins.

Gin makers in St. Mary's, Georgia, in the southeast corner of the state near the Florida border, monopolized production of Eve gins after the patent expired in 1817. Couper and Butler patronized John Pottle (also spelled Pottel). Born in Maryland, he worked in St. Mary's from the 1810s through 1850. He simplified Eve's double gin by combining two single gins back to back instead of mounting two pairs of rollers and feeders on one face. Pottle "studie[d] cheapness and simplicity," according to Couper, but his gins were "highly esteemed."[20] George Jones Kollock purchased one for his new plantation near Savannah for seventy dollars in 1837. The enslaved African mechanics on his plantation turned replacement rollers for it each ginning season.[21] J. Cart Glover was "convinced that no cotton planter ought to be without" one, and paid sixty-five dollars for it in 1845.[22]

Inclined toward the mechanical, Glover had received the gin without instructions but managed to assemble it after making a few structural adjustments. He built a gin house thirty-six feet square, used "such running gear as suits the saw gin," and powered it with "two good mules." A "prime intelligent fellow" supplied the gin and his productivity surpassed that of "prime" foot ginners, who turned out an average of twenty-five pounds of fiber per day. Glover's ginner ginned "from 250 to 335 pounds of clean cotton, requiring but little moating." His neighbors agreed with him that the staple was "less injured than by the foot gin."[23] It appeared that with Eve's self-feeding gin, Glover had achieved the elusive goal of increasing quantity without diminishing quality, but others failed.

Planter R. Furman had invested in an Eve-type gin made by a St. Mary's gin maker by the name of Farris. His neighbor had bought one from "Logan," also of St. Mary's. Even "under the most strenuous efforts and vigilant superintendance," they turned out only "60 or 70 weight per diem—often not so much," Furman complained in 1843.[24] Experience with Eve's gin left Glover satisfied but Furman disillusioned. The polar differences encapsulated the gin's problem. Technically sophisticated planters like Glover who used skilled enslaved workers found it accessible and productive. Those with less mechanical aptitude or patience wanted choices. Furman reported to the *Southern Agriculturist* that he had written to John Beath, a Boston gin maker then living in Brunswick, Georgia. Beath had patented a roller gin in 1839, which included a vibrating comb like Eve's and an endless apron-type feeding and doffing system like Whittemores' 1833 patent. He acknowledged Furman's request, and promised to send a gin, but a year passed and Furman had received nothing. "This invention appears to have undergone an extinction as mysterious, if not as sublime, as that of [the] lost star of Cassiopeia," Furman waxed, his alternatives as limited as before.[25]

Roller gin makers failed planters like Furman who were committed to the roller principle but needed gins that were more efficient and easier to use. Patentees shuffled into new configurations the elements that Eve, Reed, and the Whittemores introduced, but none produced a gin that increased outturn without damaging the fiber and that complemented plantation culture. Thus planters might have winced at Georgian Malcolm McAuley's 1849 patent for an updated Eve gin. Others would have welcomed Carolinian Francis L. Wilkinson's 1858 foot gin patent. Both were evidence of the anxious modernity of long-staple cotton planters. The McAuley and Wilkinson gins incorporated the latest in guards, doffers, and spiral-wound rollers but nevertheless they were medieval artifacts that invoked colonial and Old World associations. Little more than "two wooden rollers kept together by a wooden frame," the foot-roller gin remained the preeminent technology in the highest tier of the long-staple cotton market, the Eve gin in the lower ranks.

Frenzied saw gin patenting was but one of three responses to the panic of 1837. Some mechanics rejected the saw gin as hopelessly flawed. They introduced two radically new gins, the McCarthy gin and the cylinder gin, as candidates for the "perfected" gin. Both were patented in 1840 and promoted as "new roller gins" for short-staple cotton, but the McCarthy gin may have been conceived first. Its inventor, Fones McCarthy, is associated worldwide with Demopolis, Alabama, where he lived when he received the patent, but he spent most of his life in upstate New York. He had developed the theory and finalized the mechanics of the gin by 1838 and moved to Demopolis, a commercial town in Marengo County, in the west Alabama Black Belt, in the spring of 1840, so he could market the gin as a product of the South. From Demopolis, McCarthy traveled to Washington, D.C., and personally submitted his hand-written application for the "Smooth Cylinder Cotton-Gin" to the Patent Office on June 9. He stayed in the city until July 3, when the Commissioner made the award.[26] It was not merely a "new and useful Machine for Ginning Cotton," as he stated in the obligatory patent preamble, but a revolutionary new principle of fiber removal.

The problem of cotton ginning obsessed McCarthy even in Utica, a boom town on the Erie Canal at the foothills of the Adirondack Mountains. He studied roller and saw gins and patents and combined what he thought were the best principles into a new machine. In an 1854 patent extension, he explained his theory of fiber separation. McCarthy reduced the problem with the saw and roller gins to one of agency. Both gins damaged the cotton fiber because they rendered the fiber "an *active* agent" and "*stripped* [it] from the seed." He reconceptualized fiber as "a *passive* agent," one that was stripped from the seed. It was the seed that was the active agent in McCarthy's gin and, as such, received the brunt of the mechanical action. Consequently, the fiber was "less subject to wear, tear, and injury."[27] McCarthy reified his theory in a machine no one characterized as simple.

The open architecture of the "Smooth-Cylinder Cotton-Gin" contrasted with the obscuring boxed structure of the saw gin. It presented the three ginning components, a reciprocating blade, stationary guard, and "drawing-roller," in a flat configuration of table height (fig. 7.2). A shallow hopper cantilevered at the front of the gin and doffing mechanisms sat to the rear. McCarthy built a large roller that was standardized at four inches in diameter. He wound it spirally with roughened leather, then the practice for all roller gins. The roller rotated counterclockwise against a stationary guard, which he affixed over the roller and tangent to it. The Whittemores and others had used guards to prevent lapping, but not to enable fiber separation. Between 1840

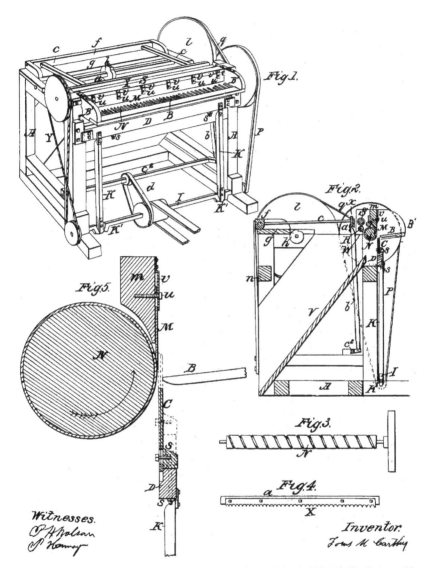

Fig. 7.2. The McCarthy Gin. Fones McCarthy, U.S. Pat. No. 1,675 (1840), Reissue No. 262, 1854; Extended 1854. Fig. 5 from the reissued patent best illustrates the McCarthy principle. The ginning cylinder (N) rotates in a counterclockwise direction as the reciprocating knife (C) moves rapidly up and down against it. The stationary knife (M) acts like a "guard," barely touching the cylinder's rough surface and directing seed cotton between it and the reciprocating knife. When in operation, the cylinder seizes the fibers of the seed cotton placed on the feeder (B), while the reciprocating blade pushes off the seed.

and 1854, McCarthy redesigned the reciprocating plate, the gin's most crucial component. In the 1840 patent, he drew it as a toothed strip and described it as a "vertically-vibrating saw" that had a "rising-and-falling as well as a backward-and-forward movement." In his 1854 reissued patent, he called the same component a "stripping plate," and limited it to a vertical reciprocating motion.[28]

The drawing roller, stationary guard, and reciprocating plate worked together in an elegant interplay of the organic and the mechanical. The ginner loaded the seed cotton into the cantilevered hopper and pushed it against the roller. The roller seized the fiber and carried it along with the seed to the stationary guard. As the fibers slid under the guard, the reciprocating plate struck the seeds repeatedly, "pushing" them off. The seeds fell through the grid at the bottom of the hopper and a series of rollers "combed" the fiber before they discharged it from the gin. It was a roller gin that, like the saw gin, demanded vigilance, not strength. But it preserved fiber quality, defined either as length or cleanliness and, its boosters promised, it matched the saw gin in outturn. The McCarthy gin appeared as a radical alternative to beleaguered short-staple cotton planters. It offered what they had yet to receive from saw gin makers, a gin that turned out high quality fiber in quantity.[29]

There was scarcely a transition period between the patent award and adoption. In 1842 Marengo County planters pooled their resources and organized "a company" that hired an agent to help McCarthy locate a shop equipped to build the gin. In May, McCarthy and the agent traveled to Paterson, New Jersey, and found a willing proprietor able to make the complicated machine. But something happened during the summer of 1842 that caused McCarthy to reconsider his singular commitment to short-staple cotton culture. In the fall he took one of the Paterson gins to Savannah and demonstrated it before a group of long-staple cotton growers. They had experienced steeper price cuts than short-staple cotton growers yet coped with the same inflationary economy and were desperate for change. Upon witnessing the McCarthy gin, one planter claimed that it was "incomparably superior to [w]hat could be produced by other Gins." Another declared that "next to Whitney, [McCarthy] has conferred the greatest benefit on the Cotton Planter."[30]

The gin that McCarthy invented for short-staple cotton had migrated into the long-staple cotton market two years after introduction. Whitemarsh B. Seabrook believed that it signaled the reunification of the cotton economy that would once again share one ginning technology. He presented McCarthy to a consortium of South Carolina agricultural societies in 1843, and declared the gin "destined to supplant Whitney's invention." It cleaned both "green seed and black-seed cotton," he said, and the green received "three cents more per pound in the Mobile market" than saw-ginned lint. The Manchester spinner H. Houldsworth agreed. The lint was "remarkably clean," he acknowledged, and "in an excellent state for our purposes as respects openness."[31]

A year later, however, both men published equally strong retractions. Seabrook now wrote that it was his "imperative duty" to inform planters that the McCarthy gin, *"as at present constructed,"* damaged the "middle and finer qualities of Long Cotton" and fell short of outturn claims. McCarthy's Georgia agent, Hiram W. Fargo of Savannah, had advertised it as capable of turning out two hundred pounds of lint per day but planters found that outturn did not exceed eighty pounds. The gin had failed to deliver on both counts. Seabrook admired McCarthy, emphasizing that he was "an intelligent man, and a skilful and persevering artist" who would solve the gin's problems and make it "highly useful, if not necessary, to the cultivators of both species of cotton grown in this country."[32] Until such time, he warned planters away. Houldsworth did not blunt his criticism. With "no hesitation," he declared, "the staple is, in *some* degree, injured by the new Gin." It was cleaner than foot-ginned lint and resembled it in appearance, but when he spun it, he discovered that its tensile strength had been impaired. The gin was suitable only for "strong, coarse fibre," such as that of short-staple cotton but should be avoided for the long.[33] In the face of censure, the gin seemed slated for the fate that befell the Reed and Whittemore gins.

Had McCarthy followed Seabrook and Houldsworth's advice, he would have returned to the short-staple cotton growers for whom he had designed the gin. Instead, in 1847, he made another strategic move to Florida and immersed himself in long-staple cotton culture. He opened a ginnery in Quincy, West Florida, which showcased the gin and provided an income. And he contracted Messrs. Jannot & Ferguson, partners of a saw gin manufactory, to build "a large number of gins."[34] For two years, they retailed the gins for approximately one hundred and fifty dollars returning ten dollars to McCarthy for each gin sold. They also shipped gins directly to him in Micanopy in eastern Florida, where he later moved. Over time, the relationship soured and McCarthy accused the firm of using "bad materials and workmanship" and "damag[ing] the reputation of the invention." When the contract terminated in 1851, he made the gin himself and supplemented production with gins made in New Jersey. His old contractors had made ten gins for him through 1851 and early 1852, "which performed so well," according to McCarthy, "that the gins were in great demand."[35]

Instead of nurturing the relationship with the Paterson shop, which had endured for a decade, he again sought a local manufacturer. The proprietor of a foundry in Jacksonville, Florida, possibly the Jacksonville Foundry and Machine Shop, assured McCarthy that he could make the gins, only to find that his workmen could not.[36] McCarthy next traveled to Augusta, Georgia, to the "large and responsible" Augusta Machine Company. The owners operated a foundry and machine shop and promised him that they could build the gins in the quantities he needed. After he finalized the contract and the company put its mechanics to work on the order, the Savannah River flooded, washing away a dam and with it the company's waterwheel. McCarthy returned to New Jersey and left with thirty-six gins. Anxiously awaiting their arrival during the "last days of the year 1852," the planters of the "Sea Island Cotton region in Florida" bought them "at once."[37]

Still unwilling to formalize a relationship with the New Jersey firm, McCarthy ranged beyond the South in the search for a manufacturer. He "repaired to the North" and "selected Mystic, in the State of Connecticut" because it "combined the greatest number of advantages."[38] Mystic River Hardware Company, popularly called "Mystic," was the same company that would make saw gins for Benjamin Gullett in the 1870s and hire him to manage their New Orleans subsidiary. McCarthy was as pleased with their work as Gullett would be. The one hundred gins they made for him in 1853 were of "better workmanship, less liable to get out of repair, stronger, and more durable than those which were previously made," he reported. His Florida customers agreed and bought them all.

McCarthy marked 1853 as a turning point in his career. He had made incremental improvements in the gin that removed lingering "objections" and had found a reliable contractor. The gin was "now fairly introduced," he believed, "and fully recognized by the public as vastly superior to all others for preparing the finer qualities of cotton."[39] South Carolina governor Robert F. W. Allston confirmed McCarthy's assessment. In an 1854 article for *De Bow's Review*, he wrote that the "Florida Gin," as the McCarthy gin was called, was "attracting much attention; and the planters are putting them up as fast as they can procure them."[40] He appended two testimonials, one that cited the gin for its " tendency to 'nep'" the cotton, the other for its delivery of lint in "exactly in the condition we require to bring it to for our work."[41] The manufacturers disagreed on the quality of fiber but agreed that its outturn rendered it a valuable invention. It turned out "150 lbs to 200 lbs. a day" with only one horse-power and one hand to feed it, replacing "five old [foot] gins." Market opinion had come full circle. From celebrating to condemning, it now confirmed that the McCarthy gin turned out quality lint in quantity.

The validation gave McCarthy the confidence to apply for a seven-year extension of the 1840 patent in the spring of 1854. The patent commissioner reviewed his hefty file but ignored the complaints of the 302 petitioners who protested the application. They charged McCarthy with failing to supply sufficient gins "or permitting others to do so," selling badly made gins, and charging "exorbitant prices" for them. Furthermore, they accused him of failing "in every instance either to send any one to put them up or to send the slightest directions how to use them."[42] The complaints exposed the underside of McCarthy and his business. Characterized as a difficult man even by admirers, he hoarded his idea almost to its detriment. He appeared to have published no newspaper notices. Excepting Jannot & Ferguson, no community of gin makers supported gin distribution and service. Only one agent, William S. Henery of Charleston, handled it in 1860.[43] The blunt letters in support and protest that filled the file profiled a single-minded eccentric, but the commissioner also appreciated their ironies. After noting that the planters complained that the gins were badly made, expensive, and hard to install, yet demanded more, he granted the extension. McCarthy received it on June 28, 1854, five days before the patent expired.[44]

In about four years, a pattern of adoption became apparent. Planters committed to the foot gin, who had not adopted the Eve gin, did not switch to the McCarthy gin. Foot gins turned out the highest quality, "first-class" Sea Island lint that received the highest prices. High prices offset the high labor costs and rewarded commitment to quality delivered by a roller gin treadled slowly and supplied gingerly by a skilled ginner. Eve gins were steam-powered by the 1850s and necessarily compromised quality. Planters and merchant ginners used them to process second- and third-quality long-staple cotton. It was in this context that the McCarthy gin became a substitute.

Planter J. Cart Glover, who had easily assembled an Eve-type Pottle gin, ran a McCarthy gin, after "removing all the objections discovered."[45] The reinforced gin house he built to withstand the vibrations of the feeder of the Eve gin accommodated the vibrations of McCarthy's reciprocating blade. Carpenters were no longer needed to turn replacement rollers but moters continued

Fig. 7.3. Steam-Powered McCarthy Gins. In a ginnery, similar to those that housed Eve gins, men on the left supply McCarthy gins while the women mote lint on a slatted table. *Frank Leslie's Illustrated Newspaper* (15 February 1862): 200 (detail). Courtesy, Library of Congress.

to work on the first floor of the ginnery, cleaning the fiber, as they had in the past (fig. 7.3).

In 1858 Leonard Wray, a representative of Britain's Royal Society of Arts, visited South Carolina on the invitation of James Henry Hammond, and noticed the new bifurcation of the long-staple cotton industry. He visited a short-staple cotton plantation first and watched saw gins in operation. Then he traveled to a Sea Island cotton plantation, about which he remarked that "the object is to preserve the quality, and thereby maintain the established reputation of the particular plantation. Upon this reputation depends not only the price, but the very sale of the cotton." But when he visited the ginnery, he was "astonished" by what he saw. "I found that the machine chiefly used for this purpose was no other than the little, humble, and despised roller gin of India—the veritable *churka*." It was not the hand-cranked model, he wrote, but a larger machine that the ginner treadled "whilst the feeding is by hand as in India." Planters considered the low outturn of "12 lbs. to 20 lbs." reasonable, he reported. Vacillating between shock and contempt, Wray explained to his British readership that "so exceedingly sensitive however is the market for this peculiar description of cotton, and so cautious are the planters, that those who grow the very finest kinds will on no consideration deviate from the beaten track, nor employ any other machine than the little, simple, old-fashion churka, actuated by the treadle."[46]

Wray's visit to a McCarthy ginnery restored his confidence in progress. An "extensive planter" ran eight steam-powered McCarthy gins and bragged that his cotton "fetches the same prices now that it did when he used the churkas." He would sooner "abandon cotton cultivation" than use "the little fiddling roller-gins or churkas," Wray quoted. The planter's emotional as well as rational commitment to the McCarthy gin lifted a "great weight" from Wray's mind that had lingered after his foot-gin encounter. The McCarthy gin was the technology he stressed to his Royal Society readership. He gave them prices for the gin and the patent rights, and supposed that they were already being made in "Manchester by Dunlop and Co."[47] Wray was not the last British industrialist whom the gin impressed. Before the end of the century, British machinists would manufacture the majority of McCarthy gins and sell them throughout the British Empire.

American mechanics, on the other hand, hesitated to adopt the complicated gin. Only two received patents before McCarthy's extended and reissued 1840 patent expired in 1861. Both may have made gins for McCarthy. James E. Ferguson of Jannot & Ferguson had moved to Micanopy when he received his 1861 patent for a minor change in the roller and a vibrating "rake or comb." Hiram W. Brown worked in Millville, a town in southern New Jersey. He received a patent in 1858 for an equally modest change. His relationship to McCarthy is less certain, but he could have been the proprietor of or worked in the Paterson shop that had made gins for McCarthy from 1842. Brown was also linked to a network of saw gin makers. He died sometime after he surrendered the patent and before a reissue was awarded. His wife Mary Jane Brown accepted it as administratrix of the estate and assigned it to herself and to Jeremiah Johnson Jr. and Franklin H. Lummus, who had witnessed the reissue with Theodore Bourne. At the time, Lummus owned the New York Cotton Gin Factory where Israel F. Brown worked after he left Clemons, Brown in Columbus, Georgia. Lummus would buy Brown's interest in Clemons, Brown while Brown would establish the Brown Cotton Gin Company in New London, Connecticut.

The Browns' involvement in McCarthy gin development deepened when Israel Brown patented a doffer for the gin and assigned it to the group named on Hiram Brown's reissue. He was fifty-two when he received the patent, his third, in 1863. By then the McCarthy gin had become synonymous with the "roller gin" and the foot gins, once so "common," were being fast forgotten as the McCarthy gin displaced them in the top tier of the long-staple cotton mar-

Fig. 7.4. McCarthy Gin, Manufactured by Platt Brothers and Company, Oldham, England, 1881 (40" wide × 28.5" high × 30" deep). This rear view of a McCarthy gin shows the spiral-wound roller and a version of Israel F. Brown's flapping doffer. Courtesy, USDA, ARS, Southwestern Cotton Ginning Research Laboratory, Mesilla Park, New Mexico. Photograph by Billy M. Armijo, supervisory engineering technician.

ket. The "improvement" Brown patented was for "that class known as the 'roller-gin'" and consisted of a "doffer or doctor" that removed fiber from the roller "with less breaking or disruption of the fiber." It was a deceptively simple device made of flaps of "some material which will have sufficient elasticity to return to its normal position after being slightly deflected." As the flaps spun, they struck the ginning roller, in a sense smacking off the fiber. Brown's flapping doffer replaced the "supplementary apparatus" that McCarthy had introduced in the reissue, becoming a hallmark of the gin. He recommended it be made of "sheet-zinc," but British manufacturers would make it out of leather or Indian rubber (fig. 7.4).[48]

The patenting and assignments suggested that American saw gin makers were investing in McCarthy technology but also that British machinists were overtaking them. Mystic River Hardware undoubtedly made the gins in Connecticut, but there is no evidence that Jannot & Ferguson; Lummus's New York Cotton Gin Factory; Clemons, Brown; or the Brown Cotton Gin Company put their patents into production. McCarthy had stopped making gins by 1860. Only fifty-five years old and wealthy, he lived alone in St. Augustine, Florida. He identified himself as a "machinist" but did not cross the five-hundred-dollar threshold for manufacturing census inclusion. The feeder he patented in 1867 evinced his continued involvement but seemed primitive when compared to the elaborate McCarthy-type gins patented by the British machinists John Platt and William Richardson in 1862; Christopher Brakell of Platt Brothers & Company, in Oldham, England, in 1865; Edward A. Cowper of Westminster in 1866; and the 1866 patent of Frederick T. Ackland, Henry G. Mitchell, and Mustapha Mustapha of Zagazig, Egypt. British citizens and subjects held the major patents by the end of the decade and promoted the gin for short- as well as long-staple cotton.

The cylinder gin was the second roller gin response to the low short-staple cotton prices of the 1840s. It received more attention from gin makers and planters than did the McCarthy gin. It too failed in its intended market and ultimately failed as a gin, two decades after its introduction. It was initially designed as a wool burrer to solve the problem of adulterated fleece. A wool burrer is a machine that removes burrs, or tiny hooked seeds that become entangled in a sheep's fleece as it grazes, along with other contaminants. Fleece adulterated with burrs was relatively inexpensive but it had to be deburred before it could be used. In the 1830s, United States wool textile makers were faced with the decision either to buy cheap burr-contaminated European fleece or to buy expensive but burr-free U.S. fleece. They chose the first option after machinists built a burrer that delivered clean fleece that was still less expensive than the domestic product.[49]

Textile machine makers were aware of the crisis in the cotton industry and believed they could apply the concept of burr removal to seed removal. They simplified the towering machine, which resembled a carding engine equipped with different kinds of cards, in order to market it as a cotton gin. They claimed that Eli Whitney's wire-toothed cylinder was the source of inspiration for the gin application, but the gin resembled his only conceptually; none incorporated the restraining breastwork. The framework of most contained two cylinders, positioned one over the other. The lower cylinder gin was spirally wound with finely toothed garnett wire; the upper cylinder took the form of a rotary, fluted beater.

In theory, ginning occurred in two concurrent steps. The upper beater cylinder revolved close to the toothed cylinder, knocking off cotton seeds. At the same time, it pushed the fiber into the crevices of the lower garnetted, fine-toothed surface. A third rapidly rotating brush cylinder brushed the fiber out of the crevices and blew it from the gin. It would seem that seed and trash fragments were pushed into the crevices along with the fiber, but neither patentees, gin makers, nor planters identified that as a problem.

Milton D. Whipple of East Douglas in central Massachusetts pioneered in 1840 with his patented "Improved Machine for Cleaning Wool from Burs," which could be applied "to the ginning of cotton, and both of the long and the short staple kinds." The machine combined rollers, toothed feeders, and a vibrating, garnetted cylinder. A "feeding apron" similar to the Whittemore invention carried the fiber to two "feeding-rollers," which may have performed a preliminary "ginning" function. A formidable spiked cylinder intermittently raked the fiber from the rollers, opening it before passing it to a "vibrating doffer," a three-inch diameter cylinder, "the main feature" of the machine. Whipple wrapped the cylinder spirally with the garnett wire one-tenth of an inch apart, "in order that the teeth may take hold of the wool or cotton, but not too rankly." Over it, he mounted a stationary guard molded to the curvature of the cylinder and "armed" with comb-teeth. "Cleaning" occurred when the fiber passed between the stationary guard and the vibrating, rotating cylinder. Burrs or seeds were unable to pass through the narrow space and were shaken free. "And thus," he concluded, "the wool is cleaned from burrs or other foreign matter, and cotton separated from its seed without injury to the fiber."[50] Whipple acknowledged his debt to the wool picker for his intermittent toothed feeder, but the idea of using a vibrating garnetted cylinder in conjunction with a stationary guard to "clean" the fiber was original to him.

Machinists from the Midwest, New England, and the Middle Atlantic states followed with patents that expanded on Whipple's idea. Lewis G. Sturdevant, a machinist from Delaware, Ohio, patented a cylinder gin in 1841 for "the Ginning of Cotton" exclusively (fig. 7.5). In place of Whipple's stationary guard, he used a rotary beater consisting of "strips of iron [vanes] placed edgewise" that extended along the length of the cylinder. The garnetted cylinder rotated underneath it. Sturdevant explained that the vanes of the beater were spaced so as to "allow the fibers of cotton to pass between them, while the seeds are beaten back and separated from the fibers by the action of the beaters." Francis A. Calvert, a machinist from Lowell, Massachusetts, followed in

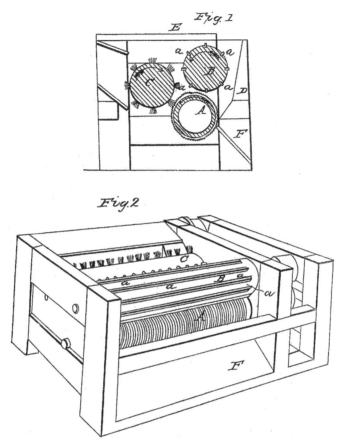

Fig. 7.5. Cylinder Gin. Lewis G. Sturdevant, U.S. Pat. No. 2,190 (1841). Sturdevant housed his gin in a saw gin frame and used beater blades (B) against a garnetted cylinder (A). A brush cylinder (C) removed the lint.

the same year with a patent reminiscent of Whipple's, and received another in 1843 that was similar to Sturdevant's, both of which he called "wool burrers and cotton gins." Theodore Ely, a New York City machinist, combined wool- and bale-breaking technology in his "Machine for Removing Burs from Wool and Seeds from Cotton," also patented in 1843.[51] Two years later, Stephen R. Parkhurst, a New York "machinist" and "woolen manufacturer" whose name became synonymous with the gin, entered the field.

In his 1845 patent, Parkhurst claimed that his "Cotton and Wool Gin" was capable of "picking, ginning, and carding wool, hemp, or cotton, so as to sepa- rate the fibers of those articles from burs, seeds, leaves, twigs, or other foreign

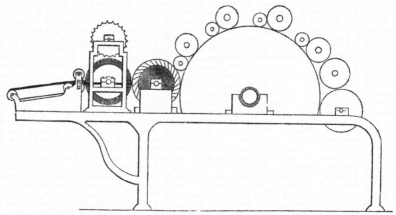

PARKHURST'S COTTON AND WOOL GIN.

PATENTED IN THE UNITED STATES, ENGLAND, SCOTLAND, FRANCE, GERMANY, &C.

The nature of this invention consists in the arrangement and combination of mechanical means, for picking, ginning and carding wool, hemp, or cotton so as to separate the fibres of those articles from burrs, seeds, leaves, twigs, and other foreign and useless substances, acting separately or in combination with the common carding machine, and, when so combined, acting in place of the tumblers usually employed with the carding machine, or when employed separately, by acting as a gin with the common whipper, or any other competent means of removing the clean material.

This engraving is here shown as acting in place of a four tumbler to a wool carding machine, when run separate from a card, or fan stripper, with card wire substituted to clean the wool from the cylinders. The cylinders are so constructed that the fibres of wool and cotton are drawn in below the outer surface, leaving the burrs and seeds, and all other worthless substances, to be separated by a metalic stripper, running close to the top cylinder—depositing the useless substances in a box directly over a feeding table to the card. By this invention, the fibres of wool and cotton are not cut nor injured, and the staple entirely preserved in its operation. It is added to a carding machine in lieu of the common tumbler, assisting the cards in carding much more wool per day and preserving the card wire; a single card carding as much wool per day as a double card, and separating all the worthless substances at the same time from the wool and cotton. When added to a worsted carding machine, more of worsted is produced from a pound of wool than by any other process now used for the manufacturing of worsted yarns. The staple being drawn straight from the feed roller and grate in the first process.

The machine can be seen on application to Naylor & Co., No. 90 John street, New-York; to whom orders should be addressed.

Fig. 7.6. Parkhurst's Cotton and Wool Gin. Stephen R. Parkhurst, U.S. Pat. No. 4,023 (1845); Reissue No. 1,137 (1861); Extended 1859. Competitors doubted whether Parkhurst's gin was capable of ginning cotton, but planters renamed it the "card gin" and embraced it. Courtesy, American Textile History Museum, Lowell, Mass.

and useless substances" and worked "separately or in combination with the common carding-machine" (fig. 7.6).[52] As in Whipple's machine, a feeding apron carried the raw fiber forward to garnett-covered "feeding rollers." A "burring cylinder" composed of toothed "steel rings" pulled the fiber under a wire grating, curved to conform to the cylinder's surface. The action pushed the fiber "below the surface of the teeth," leaving the contaminants on the surface. A revolving tin "stripper," like Sturdevant's rotary beater, then "scrape[d]" them off, burring or ginning the fiber. Parkhurst installed a brush cylinder to remove the fiber from the interstices and carried it to a main cylinder of a carding engine where it underwent conventional carding treatment.

Apparent success with the 1845 machine led Parkhurst to patent his "Parkhurst's Cotton Seed Extractor" in 1849. It was not a combination burrer but a gin that "separated cotton from the seeds without injury to the fibers." Where planters may have glossed over descriptions of "wool-burrers and cotton-gins," they would have lingered on Parkhurst's description, published in *Scientific American*, of how "bolls of cotton" were "rolled upon the [burring] cylinder, so that its teeth remove successively and gradually the outside fibers," preventing the fiber from "becoming napped or knotted." The editors received so many inquiries about the patent they felt obliged to respond. After conducting an investigation, they concluded that the "Seed Extractor" sounded like "a very valuable invention" but discovered that it had not been manufactured. They requested "the inventor or his agents [to] clear up this point," and warned planters that the silence was an admission that the patent represented "no improvement over the ordinary gin."[53]

Parkhurst fell silent because he was preoccupied with another ginning concept, this one more extraordinary. He now believed that if the teeth of ordinary carding engine cylinders were reinforced, they could gin cotton, clean the lint, and deliver it in the form of a bat reading for spinning. His 1850 "Seed-Cotton Carding and Batting Machine" did just that. A seven-cylinder monstrosity, it came equipped with one "burring cylinder," two carding cylinders, two smooth rollers that worked in conjunction with the carding cylinders, a brush cylinder, and one "batting cylinder." At a time of fluctuating but generally increasing cotton prices—and before saw gin makers popularized multi-axle gins—Mississippi cotton planters heralded Parkhurst's complicated gin. Thomas Affleck explained the response in an 1851 review of the gin for the *Southern Cultivator*. "The crisis in the affairs of the South" necessitated southerners to become "independent of every other country in manufactures," particularly in cotton textile manufacture. The "card gin," so named because cards removed "the lint from the seed . . . instead of saws," turned out "the most perfect roll of batting ready for the spinning machine or press." The lint could now travel from field to factory, dispensing "with much machinery and manipulations," facilitating the work of the planter and the textile manufacturer. "There is," he concluded, "not a doubt of the perfect working of the machine."[54] James Henry Hammond of South Carolina endorsed the gin in 1852 and Wade Hampton declared that he had "thrown away the Whitney gin, and will use no other than Parkhursts."[55]

The sectionalism that spurred the adoption of the gin did not daunt the New Yorker or his southern clients. In 1852 Parkhurst traveled to New Orleans and sold licenses to make, sell, and use the gin to southern gin makers, and personally oversaw the gin's construction. Ferdinand Smith from Daniel Pratt's gin factory followed the course of events. "All the talk" was about the Parkhurst gin, he wrote in his diary on March 9, 1852, and Pratt had already "made arrangements for making them here."[56] Four days later Parkhurst arrived at the factory from Montgomery after supervising arrangements to have his gins made at John Fraser's steam-powered gin factory. A "card gin" followed on March 18 that gave "pretty good satisfaction," according to Smith, "ginning ninety pounds in about fifteen minutes, makes a very fine sample."[57]

Building the gin challenged Smith and worried Pratt and his investors. Over the next several weeks, Pratt; E. C. Griswold, Samuel Griswold's son and the titular owner of the Prattville gin factory; and their New Orleans agent H. Kendall Carter worked in the shop with Smith overseeing his work as he tried to build a workable gin. Pratt turned to neighboring gin makers for advice. In July 1852 Smith accompanied his cousin, uncle, and Pratt to John Fraser's Montgomery factory and to the Mather & Robinson factory in Hayneville.[58] When they returned to Prattville, Ferdinand and George worked sporadically on Parkhurst's gin, but success eluded them. Parkhurst visited on April 27, 1853, and complained that the cylinder was "not right at all." Three days later, he seemed satisfied but Smith struggled with it, after he cast "altered" parts, to the end of the year. By then, Samuel Griswold had withdrawn his investment and his son from Prattville.[59]

Ostensibly, Griswold blamed Pratt for not consulting him before he diverted so much of the partnership's resources into the experimental gin, but the schism between the long-time friends betrayed deeper philosophical differences.[60] Griswold had remained a preeminent manufacturer of the "common" saw gin. Pratt publicly championed the common gin but privately pursued investment in experimental gins, like the Parkhurst gin, and in fancy saw gins, which he would manufacture for David G. Olmstead in the late 1850s. The one 1857 patent that Pratt held demonstrated his engagement in the industry's discourse. In it, he acknowledged the failure of the saw gin to deliver cotton that was not "deteriorated or injured" and the failure of the roller gin to operate efficiently. He claimed that his hopper intervention would reconcile the differences by imparting a "spiral motion" to the seed roll, allowing the saw teeth to strip the fiber from the seed without "cutting" or "breaking" it, delivering high quality fiber in quantity.[61] Rather than the conservative that Pratt's articles portrayed, the patent exposed an innovator. Griswold seemed to anticipate Pratt's ambitions and preemptively terminated the partnership of twenty years. The men publicized their decision in January 1853 in newspapers throughout the South announcing "Samuel, & E. C. Griswold" had "disposed of their interest" in their Prattville firm as of "the first day of January 1853," and that the business would henceforth "be conducted under the name and style of Daniel Pratt & Co."[62]

Even as news of the schism faded, Parkhurst abandoned the idea of the "card gin" and returned to the original cylinder gin concept, one now promoted by other northern machinists. In 1858 Parkhurst patented a cylinder gin with three cylinders, a lower "ginning-cylinder," an upper "stripper" cylinder, and a brush cylinder, essentially copying Sturdevant's expired 1841 patent. He claimed it was "compact, cheap, reliable in operation, safe for the attendants, and not liable to injury."[63] The patent drawing included a man of African descent tending the gin as he would a saw gin, a visual device Parkhurst used to reinforce the idea that it was intended as a substitute for the saw gin. It may not have succeeded in quelling concerns about Parkhurst, but planters remained interested in the cylinder gin. In the late 1850s, Alfred Jenks & Son ran a series of illustrated advertisements in *De Bow's Review* for their "Louisiana Cylinder Gin" (fig. 7.7). Where Parkhurst sold licenses to make, the Jenkses sold gins. The company was based in Bridesburgh, a Philadelphia suburb, and had supplied cotton textile manufacturing machinery to southern industrialists like William Gregg for decades.[64] Now the company appealed "To Cotton Planters."[65]

"Anything can be placed in the Breast of this Gin, such as Sticks, Trash, Bolls, &c.," the Jenkses assured planters, "as the Roller receives and takes forward nothing but the Lint, and rejects all extraneous matter." In fact, "Sand and Dirt, instead of dulling the teeth in the Roller, sharpens them." It was a gin "peculiarly" adapted to the "wants of planters who are short-handed, and gather their Cotton trashy."[66] They promised an outturn of "500 pounds of Lint in two hours" from a gin only "five and a half by three feet . . . driven with three-mule power, *easily*." They had installed the gins in a Vicksburg, Mississippi, ginnery and invited planters to witness it themselves.[67]

Martin W. Philips lived near Vicksburg and may have visited the ginnery. In August 1859 he had bought one cylinder gin and announced to readers of the *Southern Cultivator* that he expected "another for trial." "Of it I can say it is an improvement," he began cautiously. "Mr. Carroll of the house of Carroll, Hoy & Co.," in the company of another "gentleman" had examined the lint for "a half an hour at least" before agreeing that it was uninjured. Reassured, Philips returned home to "turn out some five bales" but believed he could get "8 bales per day, each weighing four hundred pounds."[68] Philips's actual out-

TO COTTON PLANTERS.

We would introduce to your notice the

LOUISIANA CYLINDER GIN, FOR SHORT STAPLE COTTON.

A machine which has been long sought for. This Gin has a Roller of a peculiar construction, filled with teeth composed of "Angular Steel Wire," and placed in the Roller tangentially to its axis, so that they always present needle points with broad backs, and are so close together that nothing but Cotton can be secreted between them, leaving the Seeds and Trash upon the surface, and the Sand and Dirt, instead of dulling the teeth in the Roller, sharpens them. In connection with this Roller is a "Stationary Serrated Straightedge," which acts in concert with it (in effect). the same as the Revolving Rollers do upon the "Sea Island Cotton," combing it under the Straight-edge, and thereby STRAIGHTENING THE FIBRE, preventing ALTOGETHER the Napping of the Cotton, and in NO MANNER shortening the Staple. The Cotton is taken from the Roller with the Brush, and thrown into the Lint Room in the usual way. The machine is simple in its construction, having but two motions, the "Roller" and the "Brush," and is not so liable to get out of order, nor to take fire, as the Saw Gin, and occupies much less space, and requires *less* power than a Saw Gin of the same capacity. A Gin of the capacity of 500 pounds of Lint in two hours, occupies a space of five and a half by three feet, and can be driven with three-mule power, *easily*. Another peculiarity of this Gin, is, that it takes the Cotton from the *surface* of the Roll, and presents it to the Brush in a thin sheet, as it passes beyond the Straight-edge, enabling the Brush to mote the Cotton in a superior manner, whilst the Roll in front of the Straight-edge is carried upon the top of it, dividing the two at that point, and following a Curved Iron or Shell, is returned again to the Cylinder, forming a Roll of about eight inches diameter; the Seeds, Bolls and Trash, being retained in the Breast by an adjustable front board, and discharged at the will of the operator, the same as the Saw Gin. The Curved Iron or Shell is capable of being adjusted so as to press the Roll as hard upon the Ginning Roller as may be desired. Anything can be placed in the Breast of this Gin, such as Sticks, Trash, Bolls, &c., as the Roller receives. and takes forward *nothing* but the Lint, and rejects ALL extraneous matter. This is a novel feature in the Gin, and peculiarly adapts it to the wants of large planters who are short-handed, and gather their Cotton trashy, as it increases the value of the Cotton from 1 to 1½ cents per pound more than that ginned upon any other machine.

There is a Roller Gin that has been in the Market for several years but the Louisiana Gin is on an entire different principle, and there being no agents for this Gin, apply direct to

ALFRED JENKS & SON,

BRIDESBURG, PA.,

MANUFACTURERS OF ALL KINDS OF

COTTON AND WOOLEN MACHINES.

sep-ly.

Fig. 7.7. Louisiana Cylinder Gin, for Short-Staple Cotton. In marketing their cylinder gin, Alfred Jenks & Son appealed to planters whose harvesters collected hulls and twigs along with cotton bolls. They claimed that the "trash" improved the performance of the gin. *De Bow's Review* 26 (June 1959): n.p.

turn of two thousand pounds of lint fell one thousand pounds short of what the Jenkses guaranteed in their notice. But it was equivalent to the outturn of a sixty-five–saw gin powered by four mules.[69] Had Philips achieved his goal of eight bales per day, the cylinder gin would have outdistanced the fastest Gullett gin.

William K. Orr, the brother and agent of Alabama saw gin maker James L. Orr, predicted planters' ultimate disappointment with the saw gin substitutes. Of the cylinder gin and the McCarthy gin, he wrote in 1857, "Many efforts have been made to supercede the saw gin, by carding or roller methods. So far they have made signal failures. Not in the sample produced, but in durability, and quantity ginned."[70] These were flimsy machines in comparison to the saw gin, Orr charged, and while they turned out superior fiber, they failed even in matching—much less exceeding—the saw gin in outturn. This was certainly true of the McCarthy gin. But there were other reasons for the failure of the cylinder gin, which eminent South Carolina and Mississippi planters testified preserved staple length and exceeded the saw gin in outturn.

One reason for its failure was the litigious behavior of cylinder gin patentees. Milton D. Whipple received the first patent for the cylinder gin principle in 1840 and surrendered it for an amended, reissued patent in 1849. In 1854 he filed for an extension that extended coverage to 1861. Although he had written his patent to cover cotton ginning, he had never used it as a gin, only as a burrer. His defense of the garnett wire-wrapped cylinder, the key component of the cylinder gin, hamstrung development. The judge in the first of Whipple's three infringement suits singled out Parkhurst, declaring that his patents infringed Whipple's if they included a garnetted cylinder. Echoing Judge Johnson's 1807 decision for Miller & Whitney, the judge stated in his 1858 decision that Parkhurst had "a right to his own improvement" but could not use any part of Whipple's patent. Nor, he added, could Whipple engraft "Parkhurst's improvement" upon his machine.[71] Ruling for the plaintiff in all three cases, the judges, however justified, supported Whipple's patent rights at the expense of cylinder gin development.

For his part, Parkhurst defended the priority of his 1845 "zigzag or pointed guard," an idea that the Whittemores had used in 1833 and undoubtedly others before them and since. He extended the patent in 1859, then surrendered it for a reissue in 1861. When he sued Israel Kinsman and two other machine manufacturers in 1847 for infringing the original patent, Kinsmen countercharged him with infringing the patent of Charles G. Sargent. Sargent along with Francis A. Calvert had purchased from Whipple the right to make his burrer, but they had patented other cylinder gins, both separately and together. Nineteen years later, the case was still in the courts. In an 1866 petition, Parkhurst claimed that he "was and is the original and first inventor . . . of each and every of said improvements" of the 1845 patent, including the guard and the toothed cylinder. From Kinsman, he extended his suit to four of the largest machine makers in Massachusetts, including the Lowell Manufacturing Company.[72] His actions, coupled with Whipple's, stifled cylinder gin manufacture and use.

Nevertheless, there were no absolute failures in the gin industry. The mill-like machines that roller gin makers introduced failed as substitutes for the foot and Eve gins on long-staple cotton. But they laid the foundations for the invention of the McCarthy and cylinder gins, both of which succeeded, but in ways their inventors could not have anticipated. Fones McCarthy's roller gin failed in its intended market but found a place in the long-staple cotton market in the United States, where it is used today. British machine makers introduced it into both long- and short-staple cotton markets throughout the Empire where it exerted the unifying effect that Whitemarsh B. Seabrook had hoped it would in the States. Even though the cylinder gin failed as a ginning technology, the garnetted cylinder returned as a component in lint-cleaning systems in the twentieth century. With the adoption of the McCarthy gin for all grades of long-staple cotton in the mid-1860s, memories of the once-ubiquitous foot gin and the vibrating Eve gin vanished; memory of the cylinder gin was even more fleeting. The saw gin withstood their challenges and retained its pride of place as the machine of the short-staple cotton culture and the emblem of the cotton South.

CHAPTER EIGHT

Machine and Myth

The cotton gin was a site of invention and innovation and a symbol of regional prosperity in antebellum America—yet it degenerated into a signifier of southern failure. Declension paralleled the extinction of the foot-roller gin and the adoption of the idea that Eli Whitney invented the first cotton gin. Phineas Miller seeded the idea in his 1794 notice, where he compared the fiber Whitney's gin turned out to that removed by fingers. Denison Olmsted developed it in his 1832 biography of Whitney, where he suggested that roller gins could not process fuzzy-seed cotton and described the enslaved using only their fingers to gin before 1794. Judge Johnson legitimized it in his 1807 decision. The idea gained credence in the 1850s, after the generation who remembered that roller gins once ginned both short- and long-staple cotton had died. In the early 1860s, when the McCarthy gin had pushed both the foot gin and Eve's gin out of the long-staple cotton market, the idea of functional roller gins went with them. The Civil War transformed the saw gin into a culpable agent of social change. The postbellum environment reshaped the gin into an ideological weapon that was wielded against freed African Americans and against the South as a whole.

Richard L. Allen, editor of the *American Agriculturist,* published an account in 1858 that typified the evolving interpretation. He wrote in the *American Farm Book* that "the introduction of Whitney's cotton gin" had given "a decided movement toward the growth of American cotton." Before it, "the separation of the cotton seed from the fibre was mostly done by hand; and the process was so slow and expensive, as to prevent any successful competition with the foreign article."[1] He correctly credited Eli Whitney's gin with facilitating the expansion of cotton production but thought that before its invention ginning had taken place only "by hand" and that the lack of a gin had prevented Americans from entering the burgeoning late–eighteenth century cotton market. The once-ubiquitous foot gin was absent from the account. The phrase "by hand" did not refer to that foot-powered machine, which had replaced hand-cranked models in the mid–eighteenth century. By the publication date, the McCarthy gin had pushed the foot gin out of the market and out of American memory; finger ginning had replaced it as the technological precedent of Whitney's gin. Allen recapitulated what was becoming a normative account of southern development. It stressed discontinuity and portrayed southerners as incompetents who had thwarted American economic development.

Allen's failure narrative perpetuated ideas that Denison Olmsted had authored in his 1832 "Memoir of the Life of Eli Whitney, Esq."[2] A pioneering meteorologist, Olmsted taught chemistry at the University of North Carolina from 1817 and held the position of State Geologist and Mineralogist from 1822. He left Chapel Hill for New Haven in 1825 to chair the Department of Mathematics and Natural Philosophy at Yale. His textbooks on natural philosophy, astronomy, and meteorology filled a void in academic curricula and were widely adopted. Cotton also fascinated him; in 1826 he patented a process to convert the seed oil into lighting gas by "destructive distillation."[3] When Benjamin Silliman, editor of the respected *American Journal of Science and Arts,* chose him to write an article on Whitney, he acknowledged a respected scholar with whose work readers were familiar. He timed the publication to deflect challenges to Whitney's claims as the inventor of the cotton gin and interchangeable parts manufacturing. The biography not only commemorated Whitney's life but also defended it.

Olmsted was as aware of the article's dual purpose as he was committed to historical accuracy. He grounded the "Memoir" with primary evidence: an interview with Eli Whitney's elderly sister, letters written by Phineas Miller and Whitney, newspaper articles, and court cases. From these sources, he selected data to accomplish his ends. On the subject of the gin, for example, Olmsted cast the roller gin as a "formidable competitor," not a predecessor, of Whitney's gin. Roller gins "extricated the seeds" by crushing them "between revolving cylinders," he wrote, and the fragments "remained in the cotton." Using selective omission, Olmsted constructed the roller gin as a machine that did not work. Despite this, it had its advocates, and they had made "great efforts . . . to create an impression in favor of its superiority [to Whitney's gin] in other respects."[4]

Whitney's heirs may have questioned the wisdom of Olmsted's unsparing account of the vagaries of the Miller & Whitney partnership, but it inadvertently rehabilitated the roller gin. Olmsted quoted Phineas Miller's letter to Eli Whitney warning him that there were "other claimants for the honor of the invention of cotton gins," and the letters from "English manufacturers" condemning the fiber produced by the wire-toothed gin. He also used excerpts of the letter written by Miller in a moment of despair when he told Whitney that "everyone is afraid of the cotton," that planters were erecting "the Roller Gins," and that merchants preferred "their cotton to ours." Including the sensitive evidence was possible because Olmsted would have Whitney triumph over these setbacks, as he ultimately did. Yet the inclusion demonstrated that roller gins were not as incapable as he had described. Why, then, would planters, after using Whitney's gin, prefer them? Why would Miller, even for a moment, himself become an advocate?[5]

The roller gin was a competitor and a substitute, Olmsted conceded, but not a precedent. Before 1794, prominent planters of South Carolina and Georgia planted cotton but had "no means of cleaning the green seed cotton," he wrote. Until such time as a machine was invented for the purpose, they could not consider raising cotton for market. Current ginning practice was simply too slow. According to Olmsted, "Separating one pound of the clean staple from the seed was a day's work for a woman; but the time usually devoted to picking cotton was the evening, after the labor of the field was over. Then the slaves, men, women and children, were collected in circles with one, whose duty it was to rouse the dozing and quicken the indolent."[6] In other words, in the narrative, sleepy and lazy enslaved Africans spent their evenings gathered in a circle "picking" cotton fiber from the seed, finger-ginning it.

Olmsted claimed that this was the level of ginning technology that Whitney saw when he debarked in Savannah in the fall of 1792. Planters complained but waited "until ingenuity could devise some machine" to process it for market. When a group of them visited Catharine Greene at her Mulberry Grove plantation, they met her guest Eli Whitney, to whom they expressed their exasperation. Retreating to a basement workshop with cotton samples procured in Savannah, Whitney made tools, drew wire, and built a machine in a matter of months "so nearly completed as to leave no doubt as to its success." Paraphrasing Judge Johnson's 1807 decision, Olmsted wrote that it "opened suddenly to the planters boundless sources of wealth, and rendered the occupations of the slaves less unhealthy and laborious than they had been before."[7] Whitney's gin, by Olmsted's reckoning, solved the planters' dilemma and relieved the enslaved from the tedium of finger ginning.

Adding the details of Joseph Eve's self-feeding gin or those of the pervasive foot gins, which he undoubtedly saw in upcountry North Carolina during the late 1810s and early 1820s, would have diluted a story made dramatic by juxtaposing the toothed gin with finger ginning. Yet the omission carried with it an insidious implication. Olmsted may not have meant to imply that southern planters were as "indolent" as the enslaved finger ginners he portrayed, but the context forced the comparison. Cotton producers all, southerners begged help, according to his account, from a tutor who admitted that he "had never seen either cotton or cotton seed in his life." Those who knew cotton best depended on one who knew it least for a machine that was indispensable to their prosperity. Olmsted's characterizations elevated Whitney at the expense of planters and enslaved Africans, reinforcing the stereotypes of the ingenious Yankee and the incompetent southerner.

Southerners, as well as northerners, grappled with Olmsted's implications. At the time of publication, most rejected the notion that the roller gin simply crushed seeds, others that it could not gin short-staple cotton. They skirted the issue and instead revived the controversies of the early 1800s, challenging Whitney's claim either to the toothed principle or to the saw gin specifically. In their response, southerners trod a precarious path. On the one hand, disavowing the roller gin as a precedent for Whitney's gin confirmed Olmsted's portrayal; on the other, championing it smacked of antiquarianism. Resolution lay in admitting, as scholars did up to the Civil War era, that the roller gin had once ginned both species of cotton, but that for short-staple cotton it had been superseded by the saw gin, which they agreed ginned faster but not better.

Thus William Scarbrough memorialized the roller gin before launching a qualified celebration of the saw gin in his "Sketch of the Life of the late Eli Whitney." Though dated to January 1832, the month and year Silliman published Olmsted's "Memoir," the article was published by the *Southern Agriculturist* in August. Scarbrough wrote it hoping to "rescue [Eli Whitney's] memory from unmerited oblivion and ingratitude." Ignoring the finger-ginning account if he was aware of it, Scarbrough wrote "It is true that [fuzzy-seed cotton] *may* be separated from its seed by rollers in the same manner that the black-seed or sea-island is," but it was more "tedious and troublesome" than using the saw gin, without which, he believed, cotton production would have faltered.[8]

Edward Baines seemed to respond directly to Olmsted's account in his 1835 history of the British cotton industry. He found the idea of finger ginning preposterous because "a man could not clean more than one pound in a day. All nations at any remove from barbarism, therefore, employ some kind of machinery," like Eve's gin and the "rude" roller gin of India. Whitney's saw gin "injures in some degree the fibre of the cotton," he wrote, but "it affords so cheap a way of cleansing it, that all the North American cotton, except the Sea Island, undergoes this operation."[9] The American historian George S. White followed a similar approach in his 1836 *Memoir of Samuel Slater.* He excerpted Olmsted but replaced the finger-ginning story with letters by Thomas Spalding and Whitemarsh B. Seabrook on the roller gin before explaining the costs and benefits of the saw gin.[10]

Accounts of cotton gin development varied during the 1840s, but the majority maintained the legitimacy of the roller gin. The editors of the *Albany New York State Mechanic* were an exception. They included a summary of the "Memoir" in their feature on Eli Whitney as an "Eminent Mechanic," in an 1841 issue. It focused only on the saw gin and mentioned neither the roller gin nor finger ginning. But Mississippi State Geologist and cotton planter Benjamin L. C. Wailes acknowledged all three in his 1841 *Address* before the Washington, Mississippi, Agricultural Society. For him finger ginning was but a transient stage in ginning development. The roller gin was the first cotton gin but using it to remove the fiber from "green seed cotton" was, he concurred, a "tedious and fatiguing process."[11] Seabrook lent credence to the idea of finger ginning in his 1844 *Memoir* by retelling a story of Virginia textile manufacturers during the American Revolution whose "field laborers" finger ginned "at the rate of 4 lbs. per week."[12] His descriptions of functioning roller gins, however, overwhelmed the solitary anecdote.

Complex explanations of the origins of the cotton gin continued into the 1850s, some contributed by northerners. In 1852, for example, New York City publisher D. Appleton in his *Dictionary of Machines [and] Mechanics* identified the roller gin as the "most ancient" gin, before he acknowledged Eli Whitney and the saw gin and its "wondrous effects on the extension of cotton cultivation."[13] Even Benjamin Silliman Jr. agreed that "mechanical means" were used to gin Sea Island cotton but contended in his catalog of the 1853 "New York Exhibition of the Industry of All Nations" that "hand labor" alone had been able to extricate the seeds of short-staple cotton.[14]

Southerners were not alone in their belief that the roller gin preceded Eli Whitney's gin but they alone risked derision if they emphasized it. Accounts and representations of the roller gin during the 1860s reflected ambivalence rather than an awareness of the implications. Bristling from the aftermath of the recently ratified Thirteenth Amendment, medical doctor William Barbee of De Soto County, Mississippi, used the gin as a measure of modernity. In his 1866 book, *The Cotton Question,* he claimed to have seen "negroes" in "old Virginia" finger ginning in 1826 but also recalled "one man in the neighborhood who had a roller-gin stand, and he was considered ahead of everybody else." Barbee thought these Virginians backward because "Whitney's gin had not yet been introduced," but he verified the functionality of the roller gin for short-staple cotton and its persistence after the introduction of the toothed gin.[15] An ulterior motive confused the account. By presenting an eyewitness account of finger ginning, Barbee meant to demean the newly freed men and women by characterizing them as having been deprived of the barest mechanical skills. Furthermore, he had probably seen moters, not finger ginners. They typically stood at moting tables (see fig. 7.3), but were also documented in drawings and photographs seated on the ground in a circle in or outside the gin house picking out seed fragments, not seeds, from roller-ginned fiber.[16]

William Ludwell Sheppard, a Virginia artist famous for his Civil War paintings, illustrated southerners' ambivalence toward the roller gin in an 1869 image entitled "The First Cotton-Gin," engraved for *Harper's Weekly* (fig. 8.1). It did not accompany an article; editors buried a short explanatory paragraph among notices and advertisements on the following page. The editors identified the image as "the primitive cotton-gin, which preceded the saw-gin invented by Eli Whitney toward the close of the last century," and accurately explained that it ginned well but did not clean the fiber as "thoroughly" as did "Whitney's machine." That Sheppard pictured a hand- rather than a foot-powered gin did not detract from the power of the image or the explanation.

Fig. 8.1. William L. Sheppard, "The First Cotton-Gin." *Harper's Weekly* 13 (December 1869): 813. Courtesy, The Historic New Orleans Collection, Acc. No. 1974.5.13.271.

Whitney's was not the first gin, the editors affirmed, but rather this "primitive" gin. This was the gin that had relieved the enslaved of "the laborious process of cleaning the cotton by hand."[17]

The image had the potential to rehabilitate planters and the enslaved without detracting from Whitney's accomplishment, but instead it reinforced social anxieties. The editors of *Harper's Weekly* demonstrated that planters had indeed used a machine to remove fiber from short-staple cotton, and the image showed enslaved Africans using it. Although it exonerated planters, the image nevertheless impugned the ginner and the roller gin. Sheppard drew two men in the foreground flanking the large hand-cranked roller gin, which the editors described as "primitive." One wears overalls and a shirt, a hat, and shoes. He catches the fiber cascading from the rollers as he watches the other man. The other man is the ginner. Sheppard dressed him as a harlequin with a collarless shirt tucked into tight calf-length striped pants and drew him barefoot. He covered his head with a bandanna, which he highlighted with a halo of cotton. The ginner uses his right hand to crank the gin, his body following rhythmically. With his left hand, he grasps a handful of seed cotton and stares at it with "excited curiosity," his mouth agape.[18]

Sheppard reserved his stereotype for the ginner, the only machine user in the scene. Women are diligently carrying seed cotton to the gin and bagging and carrying away the fiber. A young girl is sitting next to a water bucket in the right foreground, leaning forward waiting to be beckoned. Two top-hatted white men tower in the background grading fiber while a bonneted girl, a daughter perhaps, looks on. None is excited; Sheppard depicted them diligently working, carrying out ordinary chores. The image restored southerners' dignity by acknowledging that they had not planted cotton and then passively waited for a northerner to invent a gin, but it maligned, if not the "common" laborer, then the skilled enslaved African by associating him, through his dress and demeanor, with the "primitive"—if not the demonic.[19] Published ten months after Congress proposed the Fifteenth Amendment guaranteeing voting rights to men regardless of "race, color, or previous condition of servitude," it lampooned the emancipated population that had irked Barbee.

Few writers, if any, linked acceptance of finger ginning as the precedent of Eli Whitney's gin with an admission of southern incompetence. Those who did may have been daunted by the challenge of presenting the roller gin, and by association the antebellum South, as other than "primitive." They focused on challenging Eli Whitney's claim to the invention of the saw gin, hoping to glean credit for its invention. Olmsted's "Memoir" and Scarbrough's "Sketch" thus prompted no denials of finger ginning or protestations of Olmsted's roller gin characterizations. Instead, the articles generated the kinds of recriminations of Whitney that the authors had hoped to stem.

Scarbrough believed that southerners owed Whitney a debt of gratitude, but one shared by "Mrs. Miller [Catharine Greene]," the wife of the beloved Revolutionary War general Nathanael Greene. Without her, he wrote, Whitney could not have completed the gin. Confounded by the problem of removing the fibers from the wire teeth, he had paced the sitting room of Mulberry Grove, the model of the gin in hand, according to Scarbrough. Noticing his consternation, "Mrs. Miller" directed him to "turn the cylinder" and "trust to a woman's wit for the cure." She then "seized the hearth brush, perfectly unconscious" that her "playfulness" sparked in Whitney the idea of the brush cylinder. Based on no evidence other than Scarbrough's "memory unassisted by note or memorandum of any kind," the incident located a woman, a Georgian by residence, at the site of the invention.[20]

Undermining Whitney's primacy, Scarbrough's article generated others from merchants and planters still smarting from perceived wrongs perpetrated by Miller & Whitney. "A Subscriber" agreed in a September 1832 article that Whitney was a "great benefactor of the South" but reminded fellow readers that he and Miller refused to sell the gins outright or even the patent rights for others to make and sell them. Instead they wanted "to secure to themselves *the monopoly* of the invention." The "scheme . . . aroused great indignation in Georgia," he continued, and "created so strong a current of public opinion against Mr. Whitney" that success was deservedly thwarted.[21]

"A Small Planter" reminded readers, as had Olmsted, that Whitney did not invent the saw gin but the wire-toothed gin. He felt certain that an Augusta, Georgia, mechanic "by the name of Lyons" was "the genuine inventor of the annular saws." The author had seen Whitney's gin at Phineas Miller's Upton's Creek ginnery in 1797, he wrote in a December 1832 article. It had "strait wire-teeth driven into a wooden cylinder." The teeth would have "answered the purpose" if only they did not "fly out in the midst of the work and occasion considerable trouble and loss of time." Unaware of Hodgen Holmes's patent, he believed that Lyons had solved the problem. Ginnery manager John Wallace even admitted to him that Miller & Whitney "had no claim on the saw-gin, only for the original [toothed] principle."[22] Even that was fundamentally flawed, "A Small Planter" told readers. Either wire or saw teeth "so much injured" the "cotton staple" that it was "much more difficult to spin and less lasting when manufactured into cloth than that which had been ginned by rollers."[23]

With the publication of his *Cotton Planter's Manual* in 1857, Georgia editor Joseph A. Turner spiked the debate with acrimony. A collection of articles on agricultural practices "embracing a History of Cotton and the Cotton Gin," the *Manual* included a paraphrase of Olmsted's "Memoir," complete with the finger-ginning anecdote.[24] Turner balanced it with an article on the roller gin and Sea Island cotton by Thomas Spalding. He also asserted the southern claim to the toothed principle by including articles that accused Whitney of wrongfully stealing the idea from southerners and others that claimed a southerner had rightfully stolen the idea from him.[25] None questioned Olmsted's finger-ginning or roller gin portrayals. They focused on claiming the saw gin, the machine that had become an emblem of southern and national prosperity. In so doing, however, they maintained complex, if contradictory, explanations of the development of the cotton gin, establishing southerners as individuals who made rational choices informed by available technology and market exigencies. Richard Allen's 1858 *American Farm Book* portended the sea change.

During the Civil War, the idea that the saw gin was the first cotton gin gained currency. With acceptance of Olmsted's explanation came the new implication that Whitney's gin bore responsibility for not only the creation of the cotton South but also its destruction. The Reverend Henry Ward Beecher articulated this interpretation in a speech he delivered in Manchester, England, in 1863. The cost of cotton-based prosperity "arising from the invention of the cotton-gin" was "moral law," which declined in proportion to the increase in slave prices, he preached. Before its invention, an enslaved person cost "from $300 to $400." When the price reached "$800 or $900 . . . there was no such thing as moral law." When it jumped to "$1000 or $1200, . . . slavery became one of the beatitudes on the Mount."[26] The statements generated cheers from his antislavery audience but would have distressed Denison Olmsted. He had intended to create an honorable legacy for Whitney. Although he had attributed to the gin the genesis of the cotton economy, Olmsted could not have anticipated that it would be blamed for the Civil War that was raging as Beecher spoke.

After the war, popular writers and academics alike increasingly assumed that Eli Whitney's was the first gin and, explicitly or implicitly, accepted its associations. For example, C. W. Grandy, owner of the Richmond, Virginia, Southern Fertilizing Company, wrote in 1876 that his "interest in Cotton" began "when Eli Whitney invented the cotton gin. . . . Previous to that time the seeds were picked out by hand; hence the impossibility of any material production beyond the immediate wants of those who grew it." Like Barbee, he begrudged the "rude" roller gin, but only for long-staple cotton.[27] Convinced of the Olmsted interpretation, John L. Hayes wrote that the saw gin was the first cotton gin in his 1879 textbook on *American Textile Machinery*.[28]

Interpretations like Hayes's grew in frequency and specificity, most of them paraphrases of Olmsted and Scarbrough. Historian Edward C. Bates, writing the "Story of the Cotton-Gin" for the *New England Magazine* in 1890, stated that the seeds of short-staple cotton "clung to the fiber with exasperating persistency" and required "a day's work for a man to clean a pound of cotton—a rate so slow as to make its extensive production impossible." Eli Whitney was "the genius" who "unlocked the imprisoned resources of the South" permitting southerners to "utilize their slaves, land, and natural advantages" in the production of cotton.[29] Annie Nathan Meyer, founder of Barnard College, revived and extended Scarbrough's account of Catharine Greene and Eli Whitney. In 1891 she published an edited volume of articles on *Woman's Work in America* in which an author argued that Greene had previously operated her father-in-law's "anchor forge" in Rhode Island and invented the gin, after moving to Georgia.[30] She allowed Whitney to "claim the patent" because of the social opprobrium that she would have attracted. Unsure of the account's accuracy, the author hedged by paraphrasing Olmsted. "By whomever invented, no other instrument has been so fruitful of consequences," she appended. Before its invention, women "plucked the seeds from the fiber," producing one pound of fiber in a day. Afterwards, "cotton was transformed at once into a valuable commercial product. . . . It instilled such new life into the almost dying institution of slavery that the cotton-gin may well be said to have been the foster-mother of slavery in America."[31]

Historian James Ford Rhodes laid blame for slavery and the Civil War squarely on Whitney's cotton gin in his 1893 *History of the United States from the Compromise of 1850*. "Justly our boast and pride," Whitney's gin had "the effect of riveting more strongly than ever the fetters of the slave" by "prevent[ing] the peaceful abolition of slavery," Rhodes wrote, believing that the institution would have collapsed after the American Revolution. For Rhodes the line between 1793, when Whitney invented the gin, and 1861 was straight. The invention of the cotton gin enabled the cotton economy, which relied on enslaved labor, disagreement over which provoked the Civil War. "Cotton fostered slavery," he wrote, and "slavery was the cause of the war between the States."[32] Had the gin not been invented, by his logic, there would have been no cotton economy, no slavery, and no Civil War. Rhodes demonized the cotton gin by investing it with agency, as had Reverend Beecher. Both men blamed it for maintaining the viability of slavery through the cotton economy and thus for the Civil War. Added to the prevailing notions that roller gins did not function or could not gin short-staple cotton and that finger ginning instead had preceded Eli Whitney's gin, Rhodes' argument, weighted by scholarly authority, neatly explained the rise and fall of the cotton South. It was an explanation that hinged on acceptance of regional and racial ineptitude.

Historian Matthew B. Hammond made the associations explicit in his introduction to the "Correspondence of Eli Whitney Relative to the Invention of the Cotton Gin," published in the *American Historical Review* in 1897. In "recent literature" published especially by "Southern writers," there is a growing "conviction at the South," he wrote, that Whitney "was not the real inventor of the saw gin, that his gin" functioned only after "subsequent improvements by other inventors," and that he was "aided in the construction of his machine." Instead of admitting that the cotton gin was "an original product of Eli Whitney's brain," they claimed that it was "only the successful combination of the discoveries and experiments of equally brilliant but less fortunate artisans who had wrested with the same problem." Answering the challenge to Whitney's legacy, Hammond lambasted "Southern men." They were aware of the "needs and existent difficulties" in separating fiber from seed, he argued, and should have "made efforts and even important contributions toward solving this problem." Instead they "left the whole problem to be worked out by a stranger," who, quoting Olmsted, "'had never seen cotton or a cotton seed in his life.'"[33]

Hammond then presented a documentary case to test the "verdict of history" and determine if the "story of the invention in the little shop on the Savannah" was "only a historical myth." Presenting a brief but detailed and accurate history of the roller gin, including Joseph Eve's gin, he declared them all incapable of ginning short-staple cotton but adequate for Sea Island cotton. Scarbrough's claim that "Mrs. Greene" inspired the brush cylinder was "so puerile that it scarcely deserves notice," and the claims that Hodgen Holmes or Lyons or Bull had really invented the gin were mistaken. Holmes's patent, which he claimed he had examined, covered only the saws; challenges to the principle were spurious. It was, he added somewhat apologetically, "quite probable that Southern mechanics had worked on the problem of the cotton gin" and "not impossible" that they had achieved a measure of success.[34] But there was no credible evidence of it. Hammond believed that the correspondence between Phineas Miller and Eli Whitney, which he had transcribed and now published, vindicated Whitney.

Among the "Southern writers" who were defaming Whitney and the saw gin were New South boosters, textile manufacturers, and McCarthy gin makers who had revived the quest for the "perfected" gin in the face of a free labor market and competition from foreign cotton growers. Asian, African, and South American producers processed their fuzzy-seed cotton varieties with McCarthy gins and had become major suppliers during the Civil War–induced "cotton famine." Boosters argued that in order for Americans to regain their dominance, they needed to adopt the global standard. In their advocacy, they constructed the saw gin as an emblem of slavery and the Old South and the McCarthy gin as free of such associations and a worthy signifier of the New South.

They erred in the McCarthy gin attribution because by the mid-1870s British manufacturers Platt Brothers & Co. Limited of Oldham and Dobson & Barlow, Limited of Bolton made the majority. Like McCarthy himself, they had failed to convince saw gin users to adopt McCarthy gins before the Civil War but intensified their efforts after it. Platt Brothers exhibited a McCarthy gin at the Centennial Exhibition in Philadelphia in 1876 and appealed directly to southerners in an accompanying brochure. It opened with the statement that the "Patent Double Macarthy Roller Cotton Gin" was "well adapted to separate any description of cotton from its seed but specially so to the class of cotton adhering to seeds having the husk or shell covered with short fibers, technically termed 'woolly seeds,' of which American uplands is a principal variety."[35] Three years later, a consortium of British manufacturers distributed a study of various gins used in their ginneries in India, authored by John Forbes Watson, who documented the single roller gin. While Watson judged Gullett's Steel Brush gin "unequalled" in producing clean lint, he concluded that the McCarthy-type gin "stands far ahead as regards rate of out-turn and economy of power" and was "unsurpassed as regards the quality of the cotton.[36] It outproduced the saw gin, Watson contended, and produced higher quality lint.

New South boosters adopted the rhetoric of perfection, which they believed was embodied in the McCarthy gin. It was among the issues that Atlanta editor Henry W. Grady raised in his 1881 article "Cotton and Its Kingdom." "Farmers of any neighborhood" had the resources to reform cotton processing, he believed. Reform would begin with improving the gin house and installing one of "the new roller gins lately invented in England, that guarantee to surpass in quantity of cotton ginned as well as quality of lint our rude and imperfect saw gins," unwittingly applying to the saw gin the same derogatory language once reserved for the roller gin. Their material welfare was at stake, he warned planters. Boston industrialist Edward Atkinson had convinced him that farmers lost "ten per cent" from careless processing and packing.[37] After Atkinson visited Egypt and saw batteries of McCarthy gins in steam-powered ginneries in Egypt in 1894, the campaign became a crusade.

In October of 1895, Atkinson traveled from Boston to Atlanta to deliver the keynote address at the semiannual meeting of the New England Cotton Manufacturers' Association, an international consortium of textile manufacturers. The members convened in the Auditorium of the Cotton States and International Exposition, where Booker T. Washington one month earlier had delivered his controversial "Atlanta Compromise" speech, which appeared to condone racial segregation. It had been just a century since "the invention of the cotton gin," Atkinson began, but the "art" the men in the hall practiced was "full of faults and bad methods from the beginning to the end, especially at the beginning or southern end of the process." The "whole theory" of ginning and subsequent processing "rests on the form and structure of the fiber," but the saw gin violated both. Fibers should remain parallel to each other, as nature had intended, and only the McCarthy roller gin maintained this alignment, Atkinson believed. "The theory of the roller gin must prevail," he exhorted. "The saw gin must be invented out of existence."[38]

The saw gin survived the assault as it had others and entered the popular and academic literature of the early twentieth century as the first cotton gin, with finger ginning as its precedent. Images depicting finger ginners clustered in a circle illustrated trade magazines, reinforcing the idea that southern African Americans, who then endured lynching and Jim Crow laws, were not machine users and certainly not machine makers.[39] But the idea of the culpable saw gin had gained momentum; in 1956 David L. Cohn penned a stunning example. One of two books he published that year, his *Life and Times of King Cotton* earned raves from his friend Hodding Carter in a *New York Times* review. "In an almost mystic fashion," Carter wrote, Cohn had given to "King Cotton . . . a human and superhuman stature," composing "the best" interpretation "this reviewer has ever come across."[40] Carter was correct in his characterization. Cohn anthropomorphized and apotheosized the gin, but then he damned it.

Recapitulating normative history, Cohn began with a paraphrase of Olmsted on finger ginning. "Negroes often sat in a circle lit by a flaming torch and drowsily picked seeds from 'vegetable wool,'" before Eli Whitney invented the gin, he wrote. Afterwards, "the slow dying of Negro slavery" halted. The gin "stimulated it anew on a high scale," prompted the "westward spread" of cotton production, "founded a cotton plantation system, . . . [and] fostered the controversy that ended in civil war." Writing after the 1954 *Brown* v. *Board of Education* Supreme Court decision outlawing school segregation, he updated the narrative to include that Whitney's gin had also "fastened on the United States a massive race problem." Its invention coincided with the British industrial revolution, he believed, and generated industrialization in the United States. But the evil it bestowed outweighed the good. "No machine is born into Original Sin," wrote Cohn. "But in this case it was as though some wicked spirit had brooded over the young Connecticut Yankee's device and ordained: 'Evil, be thou my good.'"[41] No mere machine, the gin was the spawn of Satan.

Cohn animated the public's imagination with his dramatic characterization, but at the same time historians published repudiations of determinist explanations for southern development and of the Civil War. Kenneth Stampp's *The Peculiar Institution: Slavery in the Ante-Bellum South,* published the same year as Cohn's *King Cotton,* challenged many southern myths, among them the inferiority of Africans and complacency of slaves. In the plantation accounts he examined, Stampp found evidence that southerners were profit-maximizing entrepreneurs and that the enslaved were efficient agricultural laborers and managers who endured and resisted a brutal labor system. Stampp analyzed the institution of slavery, not southern industrialization, but by imbuing white and black southerners with agency, he undermined the foundation of Olmsted-based interpretations that assumed causal linkages between the saw gin, cotton, and slavery.[42] As Americans north and south mobilized for the civil rights movement at mid-century, two interpretations of the cotton gin and the antebellum South competed. The academic interpretation emphasized reciprocal shaping of individuals and social groups, institutions, and technologies. The popular interpretation reinforced stereotypes as it provided a succinct answer to past and present dilemmas.

In late twentieth century historical monographs, the two interpretations merged into an explanation of southern development that acknowledged human agency and market demand but maintained Eli Whitney as the solitary inventor of the cotton gin. The hybrid unwittingly bolstered the legend that Olmsted formulated. Thus American schoolchildren are taught lessons on the American Revolution, the Early National period, Jacksonian America, and the Civil War era, but they remember Whitney and the gin. Melissa Glass, a staff writer of the *Opelika-Auburn [Ala.] News,* admitted in a humorous op-ed piece published in 2001 that she had forgotten what grade she learned it in—but she "would never forget that Eli Whitney invented the cotton gin." She did not know "why it stuck" and would not recognize "a cotton gin if it fell on my head."[43] The memory "stuck" in Glass, as it does in all Americans, because it triggers a vivid narrative. The narrative begins with inept planters and sleepy finger-ginning slaves and ends with battlefield dead. It celebrates Yankee ingenuity in invention and victory and insinuates southern incompetence in passivity and defeat.

The enduring narrative hinges on the roller gin. Excluding it maintains Whitney's primacy as the inventor of the cotton gin and supports explanations of southern development that rely on discontinuity and reinforce the failure narrative. Restoring the roller gin supports explanations that emphasize continuity. They inject agency into southerners' choices and situate southerners in the global networks of commodity production and exchange in which they operated. Through these networks, they appropriated both sugar and cotton production and processing technologies and transferred them from the Mediterranean to the Atlantic economy. From settlement in 1607, indentured Europeans, enslaved Africans, and free colonists made and used mills, gins, and other machines, modifying them to suit new circumstances. Restoring the roller gin to the history of the cotton South renders the development of the gin industry, the patented response to economic crises launched by saw and roller gin makers, the invention of the McCarthy and cylinder gins, and the construction of all four types of gins in steam-powered southern machine shops, remarkable but not exceptional because it assumes southern competence.

The tragedies of slavery and the Civil War stifle the impulse to celebrate southern achievement in a success narrative centered on the cotton gin. While the gin bore no causal relationship to slavery, it processed cotton, the commodity most associated with nineteenth-century American slavery. It matters little whether it was a roller or saw gin. Neither exerted causal influences but both were integral factors in the development of a slave labor–based southern economy. Yet regional and racial integrity is risked when failure is layered onto the accomplishments of individuals who successfully exploited as well as negotiated the complexities of the slave South. The southern gin shop was a place where both occurred. Owners succeeded in rallying white and enslaved African mechanics around an industry ideal by providing an environment that rewarded personal achievement, creating an innovative industry that dominated world production at mid-century. A complex artifact, a contentious idea, the cotton gin is nevertheless testimony to the mastery they achieved, and to the regional and national prosperity they enabled.

Notes

CHAPTER ONE: Cotton and the Gin to 1600

1. Gilbert R. Merrill, Alfred R. Macormac, and Herbert R. Mauersberger, *Cotton Handbook,* 2d rev. ed. (New York: Textile Book Publishers, Inc., 1949), 78–79, 87; George Watt, *The Wild and Cultivated Cotton Plants of the World* (New York: Longmans, Green, and Co., 1907), 27, 41–45; Paul A. Fryxell, *The Natural History of the Cotton Tribe* (College Station: Texas A&M University Press, [1979]), 7 (on Watt), 126, 131 (map, fig. 51), 132.

2. Watt, *Wild and Cultivated Cotton,* 268–70.

3. Vincent T. Harlow, *A History of Barbados, 1625–1685* (New York: Clarendon Press, 1926; reprint, Negro Universities Press, 1969), 21; Hilary McD. Beckles, *A History of Barbados: From Amerindian Settlement to Nation-State* (Cambridge: Cambridge University Press, 1990), 14, 73.

4. Michael Edward Moseley, *The Maritime Foundations of Andean Civilization* (Menlo Park, Calif.: Cummings Publishing Co., 1975), 21–23; Nobuko Kajitani, "The Textiles of the Andes," *Senshoku no Bi (Textile Arts)* 20 (fall 1982), 5–6. Kajitani argues for a slightly later date.

5. Fryxell, *Natural History,* 176.

6. Dieter Schlingloff, "Cotton-Manufacture in Ancient India," *Journal of the Economic and Social History of the Orient* 17 (1974): 90; idem, *Studies in the Ajanta Paintings: Identifications and Interpretations* (Delhi: Ajanta Publications, 1987), 181, 184–85, 386 (fig. 1); Andrew M. Watson, *Agricultural Innovation in the Early Islamic World: The Diffusion of Crops and Farming Techniques, 700–1100* (Cambridge: Cambridge University Press, 1983), 32. Watson cites Schlingloff, 1974.

7. J[ohn] Forbes Watson, *The Textile Manufactures and the Costumes of the People of India* (1866; reprint, Varanasi: Indological Book House, 1982), 64; idem, *Report on Cotton Gins and on the Cleaning and Quality of Indian Cotton* (London: William H. Allen & Co., 1879), 8–9. For derived descriptions and drawings, see B. Palin Dobson, *Cotton Gins and Ginneries* (Bolton, England: Tillotsons, Ltd., 1924), 28; Evan Leigh, *The Science of Modern Cotton Spinning* (Manchester, England: Palmer & Howe, 1871; 5th ed., 1882 examined), fig. 14; Harry Hammond, "The Handling and Uses of Cotton," in *The Cotton Plant: Its History, Botany, Chemistry, Culture, Enemies, and Uses* (Washington, D.C.: G.P.O., 1896), 354; Charles A. Bennett, *Roller Cotton Ginning Developments* (Dallas: Cotton Ginners' Journal and the Cotton Gin and Oil Mill Press, 1959), 1.

8. Fryxell, *Natural History,* 169; Kate Peck Kent, *Prehistoric Textiles of the Southwest* (Albuquerque: University of New Mexico Press, 1983), 27–31. Peck describes finger-ginning, beating, and the use of the single-roller gin.

9. Y. P. S. Bajaj, ed., *Cotton* (Berlin: Springer, 1998), 4; Fryxell, *Natural History,* 166–67; Watson, *Agricultural Innovation,* 31–34.

10. R. J. Forbes, *Studies in Ancient Technology,* 2d rev. ed., vol. 4 (Leiden: E. J. Brill, 1964), 46.

11. Fryxell, *Natural History,* 168.

12. Publius Vergilius Maro (Virgil), *The Georgics,* trans. Robert Wells (Manchester: Carcanet New Press, 1982), 49.

13. Christa C. Mayer Thurman and Bruce Williams, *Ancient Textiles from Nubia: Meroitic, X-Group, and Christian Fabrics from Ballana and Qustul* (Chicago: Art Institute and University of Chicago, 1979), 37; Ingrid Bergman, *Late Nubian Textiles,* Scandinavian Joint Expedition to Sudanese Nubia, vol. 8 (Stockholm: Esselte Studium, 1975), 12–13.

14. Watson, *Agricultural Innovation,* 31–34; Forbes, *Studies,* 47, 48; F. Ll. Griffith and Grace M. Crowfoot, "On the Early Use of Cotton in the Nile Valley," *Journal of Egyptian Archaeology* 20 (June 1934): 7, 9.

15. *The Geography of Strabo,* trans. Horace Leonard Jones, 8 vols. (London: William Heinemann Ltd., 1917, 1930), 7:35 (corresponds to Strabo, Book 15, Part 1, Paragraph 21). The date of Strabo's *Geography* is contested but estimated at between the first decade of the Common Era and the last decade of the previous era.

16. Kang Chao, *The Development of Cotton Textile Production in China* (Cambridge: Harvard University Press, 1977), 5–6. Watson, *Agricultural Innovation,* 38, argued the first position; Fryxell, *Natural History,* 169–70, argued the second.

17. Chao, *Cotton Textile Production,* 77. Chao mentioned two Chinese documents, which date use of the gin to the 1180s.

18. Paul Pelliot, *Notes on Marco Polo* (Paris: Imprimerie Nationale, 1959), 501.

19. L. Carrington Goodrich, "Cotton in China," *Isis* 34 (summer 1943): 408, 410; Chao, *Cotton Textile Production,* 16–18.

20. Forbes, *Studies,* 45.

21. Watson, *Agricultural Innovation,* 33, map 4.

22. Andrew M. Watson, "The Rise and Spread of Old World Cotton," in Veronika Gervers, ed., *Studies in Textile History: In Memory of Harold B. Burnham* (Toronto, Ontario: Royal Ontario Museum, 1977), 361, fig. 3.

23. Al-Bakri, in J. F. P. Hopkins, trans., and Nehemia Levtzion and J. F. P. Hopkins, eds., annotators, *Corpus of Early Arabic Sources for West African History* (Cambridge: Cambridge University Press, 1981), 77–78, 80.

24. Al-Idrisi, in Hopkins, *Corpus,* 107.

25. Al-Dimashqi, in Hopkins, *Corpus,* 210. The editors believed that Al-Idrisi gleaned information from Al-Bakri and Al-Dimashqi from Al-Idrisi but that the authors used reliable sources.

26. William Finch, "Remembrances touching Sierra Leona, in August 1607: The Bay, Countrey, inhabitants, Rites, Fruits and Commodities," in Samuel Purchas, *Hakluytus Posthumus or Purchas His Pilgrimes* (1625; Glasgow, Scotland: J. MacLehose and Sons, 1905–7), IV:6.

27. Anne Raffenel, *Nouveau Voyage au pays des Negres* (Paris, 1856), 1:407, in Richard Roberts, "French Colonialism, Imported Technology, and the Handicraft Textile Industry in the Western Sudan, 1898–1918," *Journal of Economic History* 47 (June 1987): 463–64.

28. C. K. Meek, *The Northern Tribes of Nigeria* (London: Oxford University Press, 1925; reprint, New York: Negro Universities Press, 1961), 165–69; Pascal J. Imperato, "Bamana and Maninka Covers and Blankets," *African Arts* 7 (1974): 3, 56–57; Joanne

Eicher, *Nigerian Handcrafted Textiles* (Ile-Ife, Nigeria: University of Ife Press, 1976), 13, 16; John Picton and John Mack, *African Textiles* (London: British Museum Publications, 1979), 29; Brigitte Menzel, *Textilien aus Westafrika* (Berlin: Museum für Völkerkunde, 1972), I: plates 10–16.

29. P. Georges Mias, letter to Colonel, 6 April 1893 in Roberts, "French Colonialism," 463.

30. Hammond, *The Cotton Plant,* 354.

31. Maureen Fennell Mazzaoui, *The Italian Cotton Industry in the Later Middle Ages, 1100–1600* (Cambridge: Cambridge University Press, 1981), 2, 28–38, 50–51, 74. Mazzaoui grounded her interpretation on primary documents in Italian archives.

32. Ibid., 192 note 1.

33. Carl J. Lamm, *Cotton in Mediaeval Textiles of the Near East* (Paris: Librairie Orientaliste P. Guethner, 1937), 228.

34. Chao, *Cotton Textile Production,* 77–79; Dieter Kuhn, *Textile Technology: Spinning and Reeling* in Joseph Needham, *Science and Civilisation in China* 5:9, 191. United States cotton farmers who produced primarily for domestic purposes used gins like these through the 1930s. See John Rice Irwin, *Baskets and Basket Makers in Southern Appalachia* (Exton, Pa.: Schiffer Publishing, Ltd., 1982), 17; Allen H. Eaton, *Handicrafts of the Southern Highlands* (New York: Russell Sage Foundation, 1937), 81. Many U.S. museums, including the Smithsonian Institution and the Museum of Early Southern Decorative Arts in Winston-Salem, N.C., own models. Old Salem, Inc., has an outstanding example of a bench roller gin: Acc. No. 4378, Neg. No. S–728.

35. Sung Ying-Hsing, *T'ien-Kung K'ai-Wu: Chinese Technology in the Seventeenth Century,* trans., E-Tu Zen Sun and Shiou-Chuan Sun (University Park: Pennsylvania State University Press, 1966), 60–61. Diagrammed in Needham, *Science and Civilisation* 4:2, 123.

36. Kuhn in Needham, *Science and Civilisation,* 5:9, 190.

37. Chao, *Cotton Textile Production,* 79.

38. Needham, *Science and Civilisation,* 4:2, 123; Kuhn in Needham, *Science and Civilisation,* 5:9, 194.

39. Leigh, *Modern Cotton Spinning,* fig. 16.

40. George Basalla, *The Evolution of Technology* (Cambridge: Cambridge University Press, 1988), chap. VI. Basalla argued that efficiency played a lesser role than the cultural adaptability of artifacts in the process of selection.

41. Thomas T. Allsen, *Commodity and Exchange in the Mongol Empire: A Cultural History of Islamic Textiles* (Cambridge: Cambridge University Press, 1997), passim, 29, 46, 95–98, 100. Allsen argued that Mongols were not marauders but mediators who transferred and adopted textile and other technologies that reinforced their worldview. Focusing on *nasij,* or gold-brocaded silk fabrics as a unique signifier of Mongol culture, Allsen acknowledged the importance of cotton production and manufacture.

42. Henry Lee, *The Vegetable Lamb of Tartary: A Curious Fable of the Cotton Plant* (London: Sampson Low, Marston, Searle, & Rivington, 1887), 1–2; figs. 1–4.

43. Mazzaoui, *Italian Cotton Industry,* 141, 144, 146.

44. Alfred P. Wadsworth and Julia de Lacy Mann, *The Cotton Trade and Industrial Lancashire, 1600–1780* (Manchester: University Press, 1931), 14–17.

45. Mortimer Epstein, *The English Levant Company: Its Foundation and Its History to 1640* (1908; New York: B. Franklin, [1968]), 18–19; 36, 57, 61–62, 109, 129.

46. J. Theodore Bent, ed., *Early Voyages and Travels in the Levant* (1893; reprint, New York: B. Franklin, [1964]), xxii–xxiii. The Levant Company may have been required by the terms of its charter to use a certain percentage of British-made goods in exchange, as was the British East India Company.

47. Epstein, *The English Levant Company*, 18–19, 139, 259; see also Alfred C. Wood, *A History of the Levant Company* (Oxford: University Press, 1935; reprint, London: Frank Cass & Co., Ltd., 1964), 74–75.

48. For example, see "A Letter of Master Thomas Spurway, Merchant, . . . in a Letter to the Companie," 20 November 1617, in Purchas, *His Pilgrimes,* IV: 532.

49. Beverly Lemire, *Fashion's Favourite: The Cotton Trade and the Consumer in Britain, 1660–1800* (Oxford: Oxford University Press, 1991), 15. To the dismay of Levant Company merchants, British East Indian merchants also traded for cotton yarn. See Nicholas Downton, "Extracts of the Journall," in Purchas, *His Pilgrimes,* IV: 223.

50. Lemire, *Fashion's Favourite,* 3, 198. Lemire argued convincingly from the evidence that origins of the mass market and the industrial revolution are found in the import-substitution strategy initiated in the seventeenth century.

51. Richard Hakluyt, *Discourse of Western Planting,* eds., David B. Quinn and Alison M. Quinn (1584; London: Hakluyt Society, 1993), 4.

CHAPTER TWO: The Roller Gin in the Americas, 1607–1790

1. The First Virginia Charter, 10 April 1606. The full name of the "Firste Colonie" was the Companie of Adventurers and Planters of the Citty of London for the First Colonie in Virginia.

2. Capt. Christopher Newport, "The Description of the Now-Discovered River and Country of Virginia; with the Liklyhood of Ensuing Ritches, by England's Ayd and Industry," May 1607, in Edward E. Hale, ed., *Original Documents from the State-Paper Office, London, and the British Museum (Archaeologia Americana),* 4:61.

3. Susan Myra Kingsbury, ed., *The Records of the Virginia Company of London* (Washington, D.C.: G.P.O., 1906–1935), 3:115 (hereafter, *Records of the Virginia Company*).

4. Edmund S. Morgan, *American Slavery American Freedom: The Ordeal of Colonial Virginia* (New York: W. W. Norton, 1975), 44–91.

5. King James I, *A Counterblaste to Tobacco* (London: R. Barker, 1604; reprint, Amsterdam: Theatrum Orbis Terrarum; New York: Da Capo Press, 1969).

6. *Records of the Virginia Company,* 3:237, 309.

7. Ibid., 394.

8. Ibid., 641.

9. The seeds from South America were quite probably the smooth-seeded *G. barbadense.* Ethno-botanists question whether the Lords Proprietors could have transplanted this species to the mainland in the 1660s. A day length–sensitive perennial, *G. barbadense* could only thrive after a lengthy acclimatization process.

10. Richard S. Dunn, *Sugar and Slaves: The Rise of the Planter Class in the English West Indies, 1624–1713* (New York: W. W. Norton & Co., 1972), 6, 169: table 17, 170–71.

11. Wadsworth and Mann, *The Cotton Trade,* 31, 33.

12. A. S. Salley, ed., *Records in the British Public Record Office Relating to South Carolina,* 1663–90, 289; idem, *Commissions and Instructions from the Lords Proprietors of Carolina to Public Officials of South Carolina,* 1685–1715; South Carolina Historical Soci-

ety, *Collections,* I: 211; Great Britain, Public Records Office, Calendar of State Papers, Colonial Series, America and West Indies, 1699, 106, in Lewis Cecil Gray, *History of Agriculture in the Southern United States to 1860* (Washington, D.C.: Carnegie Institution, 1933), 54.

13. J. H. Easterby, ed., *The Journal of the Commons House of Assembly* (Columbia, S.C.: Archives Dept., 1958), 90: 6 December 1746.

14. John Lawson, *A New Voyage to Carolina* (London, 1709; reprint, Chapel Hill: University of North Carolina Press, 1967), 90, 166–67.

15. Wadsworth and Mann, *The Cotton Trade,* 15.

16. Ibid., 30–33, 72, App. G. A bag was assumed to weigh 200 pounds.

17. Ephraim Chambers, *Cyclopaedia: Or, an Universal Dictionary of Arts and Sciences* (1728; reprint, London: T. Longman, 1795–97), 335. Italics in original.

18. Ibid.

19. Gregory A. Waselkov and Bonnie L. Gums, *Plantation Archaeology at Rivière aux Chiens, ca. 1725–1848* (Mobile, Ala.: University of South Alabama, Center for Archaeological Studies, 2000), 66.

20. Edme Gatien Salmon to Count Maurepas, 12 February 1733, coll., ed., trans., Dunbar Rowland and A. G. Sanders, rev., ed., Patricia Kay Galloway, *Mississippi Provincial Archives, French Dominion, 1729–1748* (Baton Rouge: Louisiana State University Press, 1984), 4:128–29.

21. Nancy Maria Surrey, *The Commerce of Louisiana during the French Regime, 1699–1763* (New York, 1916), 177, 184, 192, 196, 209; Daniel H. Thomas, "Pre-Whitney Cotton Gins in French Louisiana," *Journal of Southern History* 31 (May 1965): 136, 140–42.

22. Mezy Le Normant to Count Maurepas, 15 December 1746, Colonies F/3 86 (drs. 229, 231, 235), Ministère de la Culture et de la Communication, Centre des Archives d'Outre Mer, Aix-en-Provence, Paris; Minister to M. Le Normant, 28 December 1750, in Thomas, "Pre-Whitney Gins," 144. I am indebted to Dr. Thomas for alerting me to the existence of these letters and drawings.

23. Denis Diderot, *Encyclopédie, ou, Dictionnaire Raisonné des Sciences, des Arts et des Métiers* (Paris: Chez Briasson, David, Le Breton, Durant, 1751; idem, *Recueil de Planches, sur Les Sciences, Les Arts Libéraux, et les Arts Méchaniques, avec leur Explication* (Paris: Briasson, David, Le Breton, Durand, 1762), 1: Oeconomie Rustique, Culture et Arsonnage du Coton (plantation and bowing scenes), Travail et Emploi du Coton (drawings of two types of gins). Originals at the Dibner Library of the History of Science and Technology, Smithsonian Institution, Washington, D.C.

24. Diderot, *Recueil de Planches,* 9–10.

25. Easterby, *Journal,* 246–47: 19 May 1747.

26. James Bromley Eames, *The English in China: Being an Account of the Intercourse and Relations Between England and China from the Year 1600 to the year 1843* (London: Pitman, 1909; reprint, New York: Barnes & Noble Books, 1974), 12, 23, 28. The French traded in Canton from 1698.

27. Hosea B. Morse, *The Chronicles of the East India Company, trading to China, 1635–1834* (1925; reprint, Taipei: Ch'eng-Wen Publishing Co., 1966), ix, 35, 40, 112–13, 135, 254–56, 264–65, 271, 275, 282 (table), 283, 291–92; Albert Feuerwerker, *State and Society in Eighteenth-Century China: The Ch'ing Empire in Its Glory* (Ann Arbor, Mich.: Center for Chinese Studies, University of Michigan, 1976), 85–86.

28. Wadsworth and Mann, *The Cotton Trade,* App. G.

29. *Charkha* is the name for the Indian spinning wheel; the proper word for gin is *belna*. Among English speakers, charkha has come to represent both machines.

30. Charles Singer, E. J. Holmyard, A. R. Hall, and Trevor I. Williams, eds., *A History of Technology,* vol. 2 (1956; reprint, Oxford: Clarendon Press, 1957), 203–4; W. English, *The Textile Industry* (London: Longmans, 1969), 1–4; Richard L. Hills, "Hargreaves, Arkwright and Crompton. Why Three Inventors?" *Textile History* 10 (1979): 114–26. Hills provides the clearest explanation of the mechanization of spinning. See also Kuhn in Needham, *Science and Civilisation,* 5:9, 215–24 on Chinese multiple-spindle, treadle spinning wheels from the fourteenth century. See Wadsworth and Mann, *The Cotton Trade,* Apps. C and D for a discussion of the invention of drafting rollers.

31. Harold Catling, *The Spinning Mule* (Newton Abbot: David & Charles, 1970), 21, 27, 32–33, 48–49.

32. British Museum, *Additional Manuscripts,* 15485, 4, in Gray, *History of Agriculture,* 183–84.

33. Robert R. Rea, "British West Florida Trade and Commerce in the Customs Records," *Alabama Review* 37 (April 1984): 157.

34. Francis Harper, ed., *The Travels of William Bartram* (Athens: University of Georgia Press, 1998), 43, 267, 272.

35. Bernard Romans, *A Concise Natural History of East and West Florida* (New York: Printed for the Author, 1775), 140–42; Robin F. A. Fabel, *The Economy of British West Florida, 1763–1783* (Tuscaloosa: University of Alabama Press, 1988), 15–17. See also Kathryn E. H. Braund, ed., *A Concise Natural History of East and West Florida,* by Bernard Romans (Tuscaloosa: University of Alabama Press, 1988), chaps. 1, 2.

36. Romans, *A Concise Natural History,* 139–40.

37. Ibid.

38. Ibid.

39. Thomas Jefferson, "Manufactures," *Notes on the State of Virginia,* in Merrill D. Peterson, ed., *The Portable Thomas Jefferson* (New York: Penguin, 1975), 216–17. *Notes* captures Jefferson's political economy at this point in time. His ideas about industry would change.

40. Tench Coxe, "Address to an assembly of the friends of American manufactures," *American Museum* 2 (September 1787): 248–54.

41. *National Cyclopaedia of American Biography,* 1896 ed., s.v. "Coxe, Tench."

42. Gray, *History of Agriculture,* 679.

43. Coxe, "Address," 1787.

44. Michael M. Edwards, *The Growth of the British Cotton Trade, 1780–1815* (Manchester: University Press, 1967), table C/3, 251.

45. Edwards, *British Cotton Trade,* 76–78.

46. Ibid., 78.

47. Ibid.

48. Romans, *A Concise Natural History,* 141. See also Thomas Anburey, *Travels through the Interior Parts of America* (1789, reprint, Boston, Mass.: Houghton Mifflin, 1923), 245–46. Anburey observed enslaved Africans ginning short-staple cotton on foot gins in Virginia in 1788.

49. John Drayton, *A View of South Carolina as Respect Her Natural and Civil Concerns* (Charleston, S.C.: W. P. Young, 1802), 133.

50. Ibid.

51. Ibid.

52. James Mease, *The Domestic Encyclopedia* (Philadelphia, Pa.: William Young Birch and Abraham Small, 1803), 3:155–56, plates 2–3.

53. John Cook, "Eve's Machine for ginning Cotton," *Bahama Gazette,* 18–21 November 1794.

54. A Cotton Planter, *Bahama Gazette,* 18–21 March 1794.

55. Thomas Spalding, "Cotton: Its Introduction and Progress of its Culture, in the United States," *Southern Agriculturist* 8 (January 1835): 44. On hernias, see Kenneth M. Stampp, *The Peculiar Institution: Slavery in the Ante-Bellum South* (New York: Random House, 1956), 306. Stampp documented "hernia" as a "very common disease among Negroes."

56. Edwards, *British Cotton Trade,* 76.

57. "Extracts from the Journal of Miss Sarah Eve," *The Pennsylvania Magazine of History and Biography* 5 (1881): 20–21; Hannah Roach Card File, American Philosophical Society, on Oswell Eve's ships and destinations; Benjamin Franklin to Richard Bache, 4 January 1773, in William B. Wilcox, ed., *Papers of Benjamin Franklin,* vol. 20: 1 January–31 December 1773 (New Haven: Yale University Press, 1976), 4–5.

58. David L. Salay, "The Production of Gunpowder in Pennsylvania during the American Revolution," *Pennsylvania Magazine of History and Biography* 99 (1975): 423, 425, 429, 439–40; "Agreement between Committee of Secrecy of Congress and Others, 1776," Samuel Hazard, ed., *Pennsylvania Archives* (Philadelphia: Joseph Severns & Co., 1853), 4:696; Oswell Eve to the Honourable the Committee of Safety for the Province of Pennsylvania, The Petition of Oswell Eve, of Frankford, 22 March 1776, *American Archives* 4th ser. (Washington, D.C., 1837–53), 464; Arthur P. Van Gelder and Hugo Schlatter, *History of the Explosives Industry in America* (New York: Columbia University Press, 1927), 44; Robert A. Howard, "Gunpowder in the American Revolution," *Arms Gazette* (July 1976): 21 for a drawing from a manuscript of Oswell Eve's gun powder mill.

59. Joseph Eve, "To the Planters and Merchants of the Bahamas," *Bahama Gazette,* 29 March 1788 (dated 19 February 1788).

60. John Wells, "Nassau," *Bahama Gazette,* 21–24 December 1790.

61. Joseph Eve, "To the Planters of the Bahama-Islands," *Bahama Gazette,* 4 January 1791.

62. "For the First Folio, Description of a Cotton Gin, Invented by Joseph Eve, of Pennsylvania," *The Port Folio* n.s. 5 (March 1811): 185–86.

63. Basil Hall, *Travels in North America in the years 1827 and 1828* (Philadelphia: Corey Lea & Carey, 1829), 221–22. Hall, a British naval officer, traveled with his wife, Margaret Hunter, who published her account of the trip separately as *An Aristocratic Journey.*

64. Ibid., 221.

65. *Port Folio* n.s. 5 (March 1811): 184–86.

66. "A Cotton Planter," *Bahama Gazette* 18 March 1794. See also John Cook, "Eve's Machine for ginning Cotton," *Bahama Gazette* 18 November 1794. Cook installed one of Eve's "Horse Gins." He did not race it against foot gins.

67. Ibid.

68. *Georgia Gazette,* 3 February 1791.

69. Thomas Spalding, "Cotton Ginning," *Georgia Gazette,* 12 May 1796.

70. Edwards, *British Cotton Trade,* Table C/3, 251.

71. Ernest McP. Lander Jr., *The Textile Industry in Antebellum South Carolina* (Baton Rouge: Louisiana State University Press, 1969), 4–6.

CHAPTER THREE: The Invention of the Saw Gin, 1790–1810

1. Eli Whitney to Thomas Jefferson, Petition for a Patent, 20 June 1793, Eli Whitney Papers, Sterling Memorial Library, Manuscripts and Archives, Yale University, New Haven, Conn., microfilm (hereafter EWP).

2. Eli Whitney Jr. to Eli Whitney Sr., 11 September 1793, EWP.

3. Eli Whitney to Thomas Jefferson, 15 October 1793, in Jeannette Mirsky and Allan Nevins, *The World of Eli Whitney* (New York: Macmillan, 1952), 72–73.

4. "Cotton Gins," *American Farmer* 4 (21 February 1823): 380–81. There are only minor differences between the 1803 copy and the *American Farmer* copy of the 1794 short description.

5. Eli Whitney, Cotton Gin Patent, Long Description, U.S. Patents Restored, Unnumbered, 25 November 1803 official copy of original 14 March 1794, retroactive to 6 November 1793, Reel 1, frames 85–93, Record Group (RG) 241, National Archives and Records Administration (NARA), Washington, D.C. Whitney's estate restored the patent and drawings on 2 May 1841, following the patent office fire of 1836. Quotations are taken from the Long Description, copy dated 25 November 1803.

6. Ibid.

7. Ibid.

8. Olmsted, "Memoir of Eli Whitney," 202–4.

9. William R. Rice, *The Centenary of Leicester Academy* (Worcester, Mass.: Printed by Charles Hamilton, 1884), 16, 20; Mirsky and Nevins, *Eli Whitney,* 24–25.

10. William L. Kingsley, *Yale College: A Sketch of Its History* (New York: Henry Holt and Co., 1879), 2:495–98.

11. Eli Whitney, *An Oration on the death of Mr. Robert Grant, a member of the Senior Class, at Yale-College, 1792* (New Haven, Conn.: Printed by Thomas and Samuel Green, 1792). Eli Whitney Jr. to [indecipherable], 10 September 1792, EWP.

12. John F. and Janet A. Stegeman, *Caty: A Biography of Catharine Littlefield Greene* (Providence, R.I.: Bicentennial Foundation, 1977), 116.

13. Phineas Miller to Eli Whitney, 20 September 1792, EWP.

14. Eli Whitney Jr., to Eli Whitney Sr., 1 November 1792, EWP.

15. Edward Rutledge to Phineas Miller, 24 August 1790, Special Collections, Duke University, Durham, North Carolina, with special thanks to Tracy Ramos, Reference Intern. Rutledge signed the Declaration of Independence and served as governor of South Carolina from 1798 until his death in 1800.

16. Eli Whitney to Eli Whitney Sr., 1 November 1792; Eli Whitney to Josiah Stebbins, 1 November 1792, EWP.

17. Eli Whitney to Eli Whitney Sr., 11 April 1793; Eli Whitney to Josiah Stebbins, 11 April 1793, EWP. There is a mysterious letter in the Eli Whitney Papers from Stebbins to Whitney dated 7 November that Whitney may have answered.

18. Eli Whitney to Josiah Stebbins, 1 May 1793, EWP.

19. Phineas Miller to Thomas Jefferson, 27 May 1793. Tucker-Coleman Papers, Earl Gregg Swem Library, College of William & Mary, Williamsburg, Virginia.

20. Phineas Miller to Eli Whitney, 30 May 1793, in Olmsted, "Memoir," 211.

21. Eli Whitney to Eli Whitney Sr., 11 September 1793, EWP.

22. Eli Whitney to Josiah Stebbins, 22 December 1793, EWP.

23. Rather than build a manufactory, Miller or Whitney may have arranged to rent property belonging to Elizur Goodrich in New Haven. See Eli Whitney to Josiah Stebbins, 22 December 1793, EWP in which Whitney referred to "Our Landlord Mr. Goodrich."

24. Conveyance of a Right to Use a Whitney Gin, 6 January 1808, in Benjamin L. C. Wailes, *Report on the Agriculture and Geology of Mississippi* (Vicksburgh: E. Barksdale, 1854), 370.

25. Miller and Whitney, "A Caution," *Georgia Gazette,* 7 May 1795.

26. Phineas Miller to James Toole, 17 May 1794, EWP.

27. Phineas Miller, "Cotton Ginning," *Georgia Gazette,* 6 March 1794; *Augusta Chronicle,* 15 March 1794. The newspapers carried the same notice, which was dated 1 March 1794.

28. George Powell & Co., *Augusta Chronicle,* 10 April 1790.

29. [Philadelphia] *Independent Gazetteer, and Agricultural Repository,* 15 December 1792; [New York] *Daily Advertiser,* 24 December 1792.

30. Thomas Jefferson to William Pearce, 15 December 1792; William Pearce and Thomas Marshall to Thomas Jefferson, 31 December 1792, in John Catanzariti, ed., *The Papers of Thomas Jefferson* (Princeton, N.J.: Princeton University Press, 1990), 24:745, 805; Thomas Jefferson to Eli Whitney, 16 November 1793, EWP; Eli Whitney to Thomas Jefferson, 24 November 1793, EWP.

31. Thomas Jefferson to Eli Whitney, 16 November 1793, EWP.

32. Eli Whitney to Thomas Jefferson, 24 November 1793, EWP.

33. D. Robinson & Co., "Wanted," *Augusta Chronicle,* 30 March 1793.

34. Wm. Kennedy & Co., "Wanted, About TWENTY NEGROE BOYS," *Augusta Chronicle,* 1 March 1794 (dated 25 February 1794).

35. Wm. Kennedy & Co., "Cotton Ginning," *Augusta Chronicle,* 18 Feb 1797.

36. Phineas Miller to Thomas Russell, 20 October 1794, to Nathaniel Durkee, 25 October 1794, EWP.

37. Phineas Miller to Major Shields, 4 November 1794, EWP.

38. Phineas Miller to Obediah Crawford; George Chatfield, 17 August 1794, EWP, Letterbook 1.

39. Phineas Miller to Thomas Shields, 22 January 1794, EWP, Letterbook 1.

40. Miller and Whitney. "A Caution." *Georgia Gazette,* 7 May 1795; *[Augusta, Georgia] Southern Sentinel, and Gazette of the State,* 21 May 1795 (dated 1 May 1795).

41. Ibid.

42. Certification of Daniel Clark Sr., John O'Connor, Lewis Alston, David Bradford, and Leonard Marbury, Natchez, 24 August 1795, Archivo General de Indias, Seville, Spain (AGI), Papeles Procedentes de la Isla de Cuba (PC), legato 32; Report of Committee Clarksville, 11 September 1795, AGI, PC, legato 32, both in Jack D. L. Holmes, "Cotton Gins in the Spanish Natchez District, 1795–1800," *Journal of Mississippi History* (1969): 162–63.

43. Ibid.

44. Report of Committee Clarksville, 11 September 1795, AGI, PC, legato 32, in Holmes, "Cotton Gins in the Spanish Natchez," 162–63. Mechanics typically used the

term "ratchet wheel" or "rag" to describe what a twentieth-century mechanic would call a "circular saw." Notably, Z. Cox received a patent for a "round saw" on 14 March 1794, the same day Whitney received the wire-tooth gin patent.

45. Phineas Miller to Peter Robinson, 2 December 1797, EWP, Letterbook 1:224.

46. Holmes, "Cotton Gins in the Spanish Natchez," 161–62.

47. Eli Whitney to Phineas Miller, 25 December 1795, EWP, Reel 1. Underscored in original.

48. Phineas Miller to Eli Whitney, 31 December 1795, EWP, Letterbook 1:27–34.

49. A Respectable Merchant, *Augusta Chronicle,* 30 April 1796.

50. Phineas Miller to Eli Whitney, 1796, in Olmsted, "Memoir," 216–17.

51. Phineas Miller to Eli Whitney, 8 March 1796, EWP.

52. Phineas Miller to Eli Whitney, 2 December 1796, EWP.

53. "Minutes of the General Assembly," *Bahama Gazette,* 25 December 1795; *Bahamas, Journal of the Council in Assembly from 10th September, 1793 to 24th December, 1801* (Bahamas: Printed by The Nassau Guardian, Ltd., Printers to the Legislature, 1935), 45, 70–73, 82, 96–98, 102–4, 113, 122.

54. Pierce Butler to Joseph Eve, 25 March 1794, Pierce Butler Letterbook, 2:331, South Carolinian Library, University of South Carolina; Joseph Eve to Benjamin Rush, 24 November 1794, in Charles C. Jones Jr., *Memorial History of Augusta, Georgia* (Syracuse, N.Y.: D. Mason & Co., 1890; reprint, Spartanburg, S.C.: The Reprint Co., 1980), 145–46.

55. Phineas Miller to Edmund Randolph, 28 June 1794, EWP.

56. Thomas Spalding, "Cotton Ginning," *Georgia Gazette,* 12 May 1796.

57. Thomas Spalding, "Notice," *Georgia Gazette,* 23 June 1796.

58. Ibid.

59. Robert Watkins, *Augusta Chronicle,* 9 July 1796; U.S. Pat. Unnumbered (1796).

60. Ibid.

61. John Course, Abraham Jones, William Longstreet, George Fee, "To the public," *Augusta Chronicle,* 1 October 1796,

62. John Currie, "Currie's Cotton Gins," *Augusta Chronicle,* 10 December 1796; William Longstreet, U.S. Pat. Unnumbered (1797); John Murr[al]y, U.S. Pat. Unnumbered (1796).

63. William Longstreet et al., "Cotton Ginning," *Augusta Chronicle,* 24 December 1796.

64. John Catlett, *Augusta Chronicle,* 11 February 1797. Longstreet later powered his barrel gins with a steam engine that he patented with Issac Briggs in 1786. William Longstreet, *Augusta Chronicle,* 22 September 1792, 10 June 1797, 14 October 1797. The steam-powered ginnery, the first recorded, burned down in 1801. *Augusta Chronicle,* 26 December 1801. Longstreet may also have run one of Miller's gins at his ginnery in 1800. See Phineas Miller to James Toole, 7 November 1800, EWP, Letterbook 2:186–87.

65. Phineas Miller to Joseph Johnson, 7 November 1800, EWP, Letterbook 2:184–85. Miller identified the two names most often associated with pirating Whitney's gin, "Bull & Lyon[s]," and specified that they made gins "after our first model," meaning the wire-toothed gin.

66. Phineas Miller to Eli Whitney, 15 February 1797, in Matthew B. Hammond, "Correspondence of Eli Whitney relative to the Invention of the Cotton Gin," *American Historical Review* 3 (Oct. 1897): 104.

67. U.S. Statutes at Large, vol. 1 (Boston: Charles C. Little and James Brown, 1845), Statute II, 21 February 1793, Secs. 2, 6, 10. This was the law that applied in the Miller and Whitney cases (hereafter Patent Act of 1793).

68. On the persistence of the opinion that toothed gins damaged the lint, see especially Phineas Miller to Liverpool textile makers Messrs. Hamilton Maher & Co., 14 June 1797; Miller to South Carolina planter Cleland Kinlock, 17 September 1798; Miller to South Carolina planter John Mayrant, 29 January 1799; Phineas Miller to Joseph Johnson, 7 November 1800. On the saw-toothed gin as the source of the problem, see Miller to U.S. Senator, the Hon. Uriah Tracy, 12 December 1797; Miller to Liverpool factor Solomon D. Aquilar, 9 April 1798. On Miller's continued use of wire-toothed gins, see Miller to Peter Robinson, 2 December 1797, EWP.

69. Phineas Miller to South Carolina planter John Mayrant, 13 December 1798; Miller to manager Henry Sadler, 10 February 1799, Letterbook 2:92–94, 109–11, EWP.

70. Patent Act of 1793, Sec. 9.

71. Ibid., Sec. 10.

72. Rule Absolute, *Phineas Miller v. Hodgen Holmes,* 30 June 1800, mss., NARA, Ga. In this document is the statement that Phineas Miller had filed a deposition against Holmes on 7 February 1799.

73. Rule Absolute, *Phineas Miller v. Hodgen Holmes,* 30 June 1800; Scire facias, *Eli Whitney & Phineas Miller v. Hodgen Holmes,* 1 July 1800; Scire facias-Plea, Hodgen Holmes adv. *Whitney & Miller,* 8 February 1802, Demurrer and Plea, *Whitney & Miller v. Hodgen Holmes,* 10 March 1802; Final Judgment on Demurrer, *Miller & Whitney v. Hodgen Holmes,* 6 November 1802, Southern Circuit Court, NARA, Ga.

74. Hodgin Holmes, "Caution!" *Augusta Chronicle,* 20 November 1802; *Georgia Gazette,* 27 November 1802; *South-Carolina State Gazette, and Columbian Advertiser,* 7 December 1802 (dated 16 November 1802).

75. Phineas Miller to Paul Hamilton, Comptroller of South Carolina, 19 January 1803, EWP, Letterbook 2:215–31.

76. Eli Whitney to Josiah Stebbins, 7 March 1803, in Hammond, "Correspondence," 123.

77. Ruth Blair, comp., *Some Early Tax Digests of Georgia* (Atlanta: Dept. of Archives and History, 1926), 276; George White, *Historical Collections of Georgia* (N.Y.: Pudney & Russell, 1854; reprint, Danielsville, Ga.: Heritage Papers, 1968), 656–57; Frank P. Hudson, *A 1790 Census for Wilkes County, Georgia* (Spartanburg, S.C.: The Reprint Co., 1988), 35.

78. Eli Whitney to Josiah Stebbins, 7 March 1803, in Hammond, "Correspondence," 123.

79. Eli Whitney to Josiah Stebbins, 15 October 1803, in Hammond, "Correspondence," 124.

80. Franklin B. Dexter *Biographical Sketches of the Graduates of Yale College* 4th series (New York: Henry Holt & Co., 1911), s.v. "Josiah Stebbins."

81. Deposition, *Miller & Whitney v. Arthur Fort et al.,* Southern Circuit Court, 10 December 1803, NARA, Ga.

82. *Whitney v. Carter,* 1810, Case No. 17,583, 29 *Fed. Cas.,* 1070–73.

83. Ibid., 1072.

84. Phineas Miller to Paul Hamilton, Comptroller of South Carolina, 19 January 1803, EWP, Letterbook 2:216–17.

CHAPTER FOUR: The Transition from the Roller to the Saw Gin, 1796–1830

1. Tench Coxe, *A Statement of the Arts and Manufactures of the United States of America, for the Year 1810* (Philadelphia: Printed by A. Cornman, 1814), ix–x.

2. Ibid.

3. [Levi Woodbury], *Cotton, Cultivation, Manufacture and Foreign Trade of: Letter from the Secretary of the Treasury* (Washington, D.C.: Blair & Rives, 1836), 16–17, 24–25. Gray, *History of Agriculture,* 1027.

4. See Robert W. Fogel and Stanley L. Engerman, *Time on the Cross: The Economics of American Negro Slavery* (1974; reprint, Lanham, Md.: University Press of America, 1984), 60–61, 87–94 for a discussion of cotton prices relative to slave prices over the antebellum period.

5. Edwards, *British Cotton Trade,* 41–42.

6. Ibid., 89, 99.

7. R. S. Fitton and A. P. Wadsworth, *The Strutts and the Arkwrights, 1758–1830: A Study of the Early Factory System* (Manchester: University Press, 1958), 286, 288. Strutts's complaints about South American imports date from the 1820s. He bought 109 bales of Georgia Upland cotton in 1819; he complained about overcharges for "canvas & twine," but not fiber length.

8. Phineas Miller to Charles Cotesworth Pinckney, 14 November 1801, EWP.

9. Judgment. *Henry Keebler* v. *Eli Whitney,* survivor of Miller & Whitney, 2 January 1804, NARA, Ga. Watkins here represented Keebler and other defendants including Silus Gregg, Daniel Easley, and Jonathan Embree. Watkins invented and patented a barrel gin in 1796.

10. Phineas Miller to John Mayrant, 13 December 1798, 20 January 1799, EWP, Letterbook 2.

11. Phineas Miller to Obadiah Crawford, 10 August 1799, EWP, Letterbook 2.

12. O. Crawford, "Notice," *Augusta Chronicle,* 28 July 1815. In equivalent year 2000 dollars, Crawford's gin cost $1,336. This and subsequent price equivalencies were calculated on the Economic History Services website "How Much is That?" (www.eh.net/hmit), which is the online edition of John J. McCusker's *How Much Is That in Real Money: A Historical Price Index for Use as a Deflator of Money Values in the Economy of the United States* (Worcester, Mass.: American Antiquarian Society, 1992).

13. James Miller, Senior, "Sales 48 Cotton Gins for Account Mr. E Whitney of New Haven," Blake Family Papers, Group 85, Sterling Memorial Library, Yale University, New Haven, Conn.

14. Thomas & J. Reid, *Augusta [Ga.] Chronicle,* 22 March 1811.

15. Isaac Anthony Jr., "For Sale, Cotton Gins," *Augusta [Ga.] Chronicle,* 9 August 1811.

16. Holmes, "Cotton Gins in Spanish Natchez," 161, 166, 168; Francis Baily, *Journal of a Tour in Unsettled Parts of North America, in 1796 & 1797,* ed. Jack D. L. Holmes (Carbondale, Ill.: Southern Illinois University, 1969), 152, notes 32–33, pp. 294–95.

17. Eleazer Carver, "To the Editors," *American Polytechnic Journal* 2 (July to December 1853): 383.

18. Martin L. Thomas, "Temple of Industry," *[Natchez] Mississippi Republican,* 13 March 1821.

19. Eleazer Carver, Statement of Rects & Expenditures, Document No. 10, filed 4 October 1858, petition for the extension of U.S. Pat. No. 3,875 (1845), Extension Case Files, Box 51, Record Group (RG) 241, NA.

20. *Charleston Directory* (Charleston: J. J. Negrin, 1802), 8, 55, 63, 75; *Directory for the District of Charleston* (Charleston, S. C.: Printed by John Hoff, 1809), 37.

21. J. P. Bron in *Nelson's Charleston Directory, and Strangers Guide* (Charleston, S.C.: Printed by John Dixon Nelson, 1801); Bron, Rigalle, Symonds in *Charleston Directory, and Stranger's Guide* (Charleston, S.C.: Printed by John A. Dacqueny, 1802); Fullilove in *Directory for the District of Charleston* (Charleston, S.C.: Printed by John Hoff, 1809).

22. Benjamin Prescott, "Black Smith Shop," *[Columbia] South-Carolina State Gazette, and Columbian Advertiser,* 10 December 1808.

23. Vestry Minutes, Bahama Islands, Parish of St. Matthew [Anglican] Church, vol. 1, 1802–14, Department of Archives, Nassau, Bahamas.

24. Joseph Eve, "Cotton Machine," *Charleston Courier,* 23 September 1805 (dated 10 April 1805); idem, "Advertisement," *Charleston Courier* 4 December 1805 (dated 23 September 1805).

25. "A Cotton Gin," *City Gazette and Daily Advertiser, Charleston [S.C.],* 3 July 1800.

26. *The Port Folio* n.s. 5 (March 1811), 184–86. See chap. 2.

27. Benjamin Bethel, "Cotton Gins," *[Charleston, S.C.] Strength of the People,* 6 September 1810 (dated 14 August).

28. Joseph Eve, *Better to Be: A Poem, in Six Books* (Augusta: Printed at the Chronicle and Advertiser Office, 1823); W. A. Clark, "An Unremembered Poem," *Augusta Chronicle,* 14 July 1912. Clark published the epitaph Eve wrote when he was seventy-six years old. It began "Here rests one fortune never favored . . . "; John Bourne, *A Treatise on the Steam Engine* (London: Longman, Brown, Green, and Longmans, 1846), 185; Roland Turner and Steven L. Goulden, eds., *Great Engineers and Pioneers in Technology,* vol. 1: *From Antiquity through the Industrial Revolution* (New York: St. Martin's Press, 1981), 336–37, s.v. Joseph Eve.

29. This amount would be equivalent to the purchasing power of $139,000 in year 2000 dollars.

30. A Citizen of Camden, "M'Bride's Carding and Spinning Cotton Machines," 19 September 1809 (dated 13 September 1809); "Homespun," 29 September 1809, *[Charleston, S.C.] City Gazette and Daily Advertiser;* Lander, *The Textile Industry in Antebellum South Carolina,* 6–10.

31. Bloomfield & Elliott, "The Improved Cotton Gins and Patent Reversed Grist Mill Manufactory," *Raymond [Miss.] Times,* 13 November 1840 (dated 20 January 1840).

32. Michael P. Johnson and James L. Roark, *Black Masters: A Free Family of Color in the Old South* (New York: W. W. Norton, 1984), chap. 1. This is an unparalleled reconstruction of the exceptional life of an antebellum southern African-American saw gin maker.

33. William Ellison, "Cotton Saw Gins," *[Sumpterville, S.C.] Southern Whig,* 12 January 1833 (dated 17 May 1832).

34. South Carolina, vol. 9, 144–47, R. G. Dun & Co., Baker Library, Harvard University Graduate School of Business Administration (hereafter R. G. Dun & Co.).

35. Allen Jones & Drury Campbell, Gin Making; William Atkinson & John Workman, "Saw Gins made and Repaired (at the sign of Sheaf, Rake & Hoe)," *Business Directory, 1816–24* (Camden, S.C.), microfiche no. 263; "Cotton Saw Gin," *Southern Chronicle and Camden [S.C.] Aegis,* 28 April 1824 (dated 14 April 1824).

36. John Workman, "Cotton Gins," *Camden [S.C.] Journal*, 30 June 1841.

37. John Graham, Williamsburg District, South Carolina, manufacturing schedule, U.S. Census, 1820.

38. *Directory and Stranger's Guide, for the City of Charleston* (Charleston, S.C.: Printed by Archibald E. Miller, 1822), 56; *Directory and Stranger's Guide, for the City of Charleston* (Charleston, S.C.: Printed and Published by A. E. Miller, November 1824), 61; *Directory or Guide to the Residences and Places of business of the Inhabitants of the City of Charleston and Its Environs* (Charleston, S.C.: Printed by James S. Burges, 1828), 59; *Directory and Strangers' Guide, for the City of Charleston and its Vicinity* (Charleston, S.C.: Printed at the Office of the Irishman, 1831), 90; *The Charleston Directory* (Charleston, S.C.: Daniel J. Dowling, Publisher and Printer, 1835), 53; *Dowling's Charleston Directory* (Charleston, S.C.: P. L. Dowling, 1837), 65.

39. Jane Gill of Fulton Co., Ga., identified herself as a "cotton gin maker" in the 1860 U.S. Census. She was born in South Carolina and was sixty years old at the time. Rhode Island–born Ruth A. Sprague of Natchez, Adams Co., Miss., identified herself as a "gin wright" in the 1860 U.S. Census. She was twenty-six years old. Neither of the women owned slaves.

40. Martin Thornton, "Cotton Gins," 31 March 1818; Elisha Reid, "Saw-Gin Business," 16 June 1818; A. J. Brown, "Notice," 17 October 1818, *[Milledgeville, Ga.] Reflector*.

41. "Awful Calamity by Fire!" *[Columbus, Ga.] Muscogee Democrat—Extra*, 9 October 1846.

42. Robert Freeman, James Nickerson, Ga., Jones Co., U.S. Census, 1820.

43. Thomas Johnson, Ga., Washington Co., U.S. Census, 1820.

44. Ibid.

45. John O. Grant, Ga., Pulaski Co., U.S. Census, 1820.

46. David Moore, "Cotton," *Huntsville [Ala.] Gazette*, 21 December 1816.

47. Brittain Huckaby, "For Sale, A Tract of Land," *[Augusta, Ga.] Columbian Sentinel*, 22 November 1806.

48. Brittain Huckaby, "Dutch Fans & Spinning Machines," *[Milledgeville] Georgia Journal*, 10 May 1815.

49. Brittain Huckaby, "To Planters: Improved Cotton Gin and Mill," *[Natchez] Mississippi State Gazette*, 15 May 1824 (dated 22 March 1824), U.S. Patent Unnumbered, Unrestored (1824); David Williamson, "Notice," *[Huntsville, Ala.] Alabamian*, 15 July 1824.

50. Brittain Huckaby, "To Planters: Improved Cotton Gin and Mill," *[Natchez] Mississippi State Gazette*, 15 May 1824 (dated 22 March 1824).

51. A. W. Bell, "Gin Making and Screw Cutting," *The [Ala.] Tuscumbian*, 26 July 1826 (dated 22 June).

52. Saw Gin, 1830s. Collection Old Salem, Winston-Salem, North Carolina, Acc. No. 606, Negative No. S-725; Charles Kuralt, "Home-Made Gin Old Days Reminder," *Charlotte [N.C.] News*, n.d. The *News* published a photograph of the gin and its owner with the article. The curator and registrar of Old Salem graciously allowed me to open the gin to confirm the details.

53. A. J. Brown, "Saw-Gin Business," *[Milledgeville, Ga.] Reflector*, 16 June 1818.

54. Henry Clark, U.S. Pat. Unnumbered (1836).

55. Edwin Keith, U.S. Pat. Unnumbered (1836). Charles A. Bennett, *Saw & Toothed Cotton Ginning Developments* (Dallas: The Cotton Ginners' Journal and the Cotton Gin and Oil Mill Press, 1960), 25.

56. Eleazer Carver, Statement of Expenses, 1858, Patent Extension File, RG 241, NA.

57. *Laws of the Commonwealth of Massachusetts,* vol. 10, chap. 58 (Boston: Dutton and Wentworth, 1828), 107.

58. *Laws of the Commonwealth of Massachusetts,* vol. 9, chap. 45 (Boston: True and Greene, 1825), 257.

59. *Laws of the Commonwealth of Massachusetts,* vol. 10, chap. 20, p. 35.

60. Martin L. Thomas, "Temple of Industry," *[Natchez] Mississippi State Gazette,* 17 March 1821.

61. E. Carver, "Cotton Gin Irons & Stands," *[Natchez] Mississippi State Gazette,* 3 April 1819.

62. John Winslow, "Gin Stands and Gin Irons, Very Cheap!!" *[Natchez] Mississippi State Gazette,* 12 August 1820.

63. Jonathan Williams and James Cuddy, "Blacksmithing," *[Natchez] Mississippi State Gazette,* 4 December 1821; idem, *Mississippi Republican,* 2 January 1822; Zephaniah Burt, *[Natchez] Mississippi Republican,* 9 January 1822.

64. John Edson, "Gin and Mill Wright," *[Natchez] Mississippi State Gazette,* 27 October 1821; idem, *[Natchez] Mississippi Republican,* 2 January 1822.

65. Martin L. Thomas, *[Natchez] Mississippi Republican,* 13 March 1821.

66. *Mississippi State Gazette,* 18 December 1824.

67. Martin L. Thomas, *Port-Gibson [Miss.] Correspondent,* 14 September 1826.

68. Tobey, Thomas & Co., "Gin Stands," *Mississippi Statesman, and Natchez Gazette,* 14 February 1827.

69. "Improved Mode of Cleaning Cotton," *American Farmer* 3 (2 November 1821): 256.

70. M. B. Sellers, "Gin Stands for Sale," *Port-Gibson [Miss.] Correspondent,* 12 May 1821; E. B. Clarke, "Cotton Gins Complete," *[Natchez] Mississippi Republican,* 31 July 1821.

71. J. J. Chewning, "Aikins' Gin Stands," *[Vicksburg, Miss.] Advocate & Register,* 18 October 1832; A. Glass, "Gin Stands," *Port-Gibson [Miss.] Correspondent,* 2 August 1822; James J. Chewning, "Aikin's Improved Gin Stands," *[Vicksburg, Miss.] Advocate & Register,* 12 September 1832.

72. William A. Aikin, "Notice to Planters," *Florence [Ala.] Gazette,* 19 August 1824.

73. Winter and Moore, "Gin Machinery," *[Natchez] Mississippi Republican,* 26 December 1820; idem, Gabriel Winter, "Iron Machinery," *[Natchez] Mississippi State Gazette,* 25 May 1822.

74. Deed of Mortgage, Samuel Griswold and David White, 11 November 1819, transcript in Bruce G. and Walter W. Spengler, *Griswoldville: A Collection of Maps, Pictures, Stories, and Personal Comments* (Clinton, Ga.: The Authors, 1995), I:95.

75. Georgia, vol. 18, p. 205, R. G. Dun & Co.

76. *Report of the Secretary of State of Such Articles Manufactured in the United States As Would Be Liable to Duties If Imported from Foreign Countries . . . with a Schedule of Factories, 27 January 1824* (Washington, D.C.: Printed by Gales & Seaton, 1824), 45.

77. Griswold & Clark, "Cotton Gins, &c.," *[Milledgeville] Georgia Journal,* 7 June 1825, in Spengler, *Griswoldville,* I:2.

78. Ibid.

79. Samuel Griswold, "Cotton Gins," *[Milledgeville] Georgia Statesman,* 27 June 1826, in Spengler, *Griswoldville,* I:2.

80. The 1820 census was damaged, preventing a clear reading of enslaved males in the two age categories "14 and 26" and "26 and 45." Griswold owned two men "45 and upwards," two "free colored men" between "14 and 26" and two between "26 and 45."

81. D. Hamilton Hurd, comp., *History of New London County, Connecticut* (Philadelphia: J. W. Lewis, 1882), 241; *Biographical Review*, vol. 26: *Containing Life Sketches of Leading Citizens of New London County, Conn.* (Boston: Biographical Review Publishing Co., 1898), 72; *Genealogical and Biographical Record of New London County, Connecticut* (Chicago: J. H. Beers & Co., 1905), 140.

82. Edwards, *British Cotton Trade*, 43. Edwards argued that the shift in emphasis on quantity over quality began in 1788 and believed that price not fashion drove it. See also 30–31, 87–88.

CHAPTER FIVE: The Saw Gin Industry, 1830–1865

1. Woodbury, *Cotton, Cultivation, Manufacture*, 13; J. D. B. De Bow, *The Seventh Census of the United States: 1850* (Washington, D.C.: Robert Armstrong, 1853), table clxxxv, 173 (includes 1840 figures); Stuart Bruchey, *Cotton and the Growth of the American Economy: 1790–1860* (New York: Harcourt, Brace & World, Inc., 1967), tables 3, C–I.

2. Douglass C. North, *The Economic Growth of the United States, 1790–1860* (New York: W. W. Norton, 1961; 1966), 68–69.

3. Charles S. Aiken, "An Examination of the Role of the Eli Whitney Cotton Gin in the Origin of the United States Cotton Regions," *Proceedings of the Association of American Geographers* 3 (1970): 5. An historical geographer, Aiken demonstrates that demand, not an easier way to process cotton, drove the expansion of the cotton economy.

4. Charles M. Johnson, "Gin-Making," *Greene County [Ala.] Gazette*, 31 May 1830.

5. William Avery, "Cotton-Gin For Sale," *Greene County [Ala.] Gazette*, 23 August 1830.

6. Sturdevant & Hill, "Gin Manufactory," *Selma [Ala.] Free Press*, 6 February 1834.

7. Cicero Broome and Thadeus Mather, "Gin Manufactory," *[Hayneville] Alabamian*, 5 September 1840 (dated 1 October 1839).

8. William W. Gaines, "To Planters," *[Vicksburg] Mississippian*, 12 September 1832.

9. John Hebron Moore, *The Emergence of the Cotton Kingdom in the Old Southwest: Mississippi, 1770–1860 (Baton Rouge: Louisiana State University Press, 1988)*, 211–13.

10. George White, *Statistics of the State of Georgia* (Savannah: W. Thorne Williams, 1849), 355, 447, 532; idem, *Historical Collections of Georgia* (New York: Pudney & Russell, 1855), 505, 511, 566, 568–71.

11. Weymouth T. Jordan, *Ante-Bellum Alabama: Town and Country* (Tuscaloosa: University of Alabama Press, 1957), 141. After searching the *[Huntsville] Southern Advocate* from the late 1840s, I could not locate the original editorial from which Jordan took the excerpt. In other editorials, Figures presented himself as a booster of southern industry. The quotation may represent exasperation with planters rather than with unfamiliarity with southern manufacturers.

12. William Gregg, "Southern Patronage to Southern Imports and Domestic Industry," *De Bow's Review* 29 (July 1860): 77–83. Gregg attributes the "failure of southern manufacturing" to the "want of southern patronage."

13. Georgia vol. 18, p. 205, R. G. Dun & Co.

14. E. C. Griswold to Samuel Tippett, 19 October 1842, in Carolyn W. Williams, *History of Jones County Georgia: For One Hundred Years, Specifically 1807–1907* (Macon, Ga.: J. W. Burke Co., 1957), 469.

15. Contract Authorizing Francis S. Johnson Continued Management of the Cotton Gin Business, 23 Nov. Tuscaloosa 1842, in Spengler, *Griswoldville*, 3:n.p.

16. S. H. Griswold, "The Cotton Gin," in Williams, *History of Jones County Georgia*, 506–7.

17. Curtis J. Evans, *The Conquest of Labor: Daniel Pratt and Southern Industrialization* (Baton Rouge: Louisiana State University Press, 2001). In this definitive work on Pratt as a paradigm of southern industrialists, Evans examines his purchase, hire, and use of African mechanics as well as the relationships that white mechanics formed among themselves and within the community.

18. Williams, *History of Jones County Georgia*, 204, 207, 237. Lowther witnessed Griswold's 1819 "Deed of Mortgage."; Daniel Pratt to Edward Pratt, in Malcolm C. McMillan, *Daniel Pratt: Ante-bellum Southern Industrialist* (Prattville, Ala.: Continental Eagle Corp., 1998), 3.

19. S. F. H. Tarrant, ed., *Hon. Daniel Pratt: A Biography* (Richmond, Va.: Whittet & Shepperson, 1904), 21; Evans, *Conquest of Labor*, 13–14.

20. A. H. Burdine, Saw-Sharpener, U.S. Pat. No. 21,483 (1858); Saw-Sharpener, No. 24,790 (1859); Saw Gin Grate, No. 29,765 (1860).

21. Evans, *Conquest of Labor*, 22–23.

22. Daniel Pratt, "Gins Sold, 1836," Malcolm C. McMillan Collection, Box 1, Reel 16C, Auburn University Library Archives, Auburn, Ala. Microfilm.

23. Daniel Pratt, "Cotton Gins," *De Bow's Review* 2 (September 1846): 153. Historians dispute whether Griswold co-owned the gin factory with Pratt from its inception or leased it from Pratt from 1848 to 1853. The evidence points to a much larger stake on the part of Griswold than of Pratt. The 1850 census lists Griswold's son as owner of the gin factory and the foundry, Pratt as owner of the sawmill and gristmill only. The textile mill was owned by the Prattville Manufacturing Company.

24. *The Journals of Ferdinand Ellis Smith, The Journals of George Littlefield Smith*, ed. Larry W. Nobles (Prattville, Ala.: Autauga Genealogical Society, 1991), 1. Hereafter F. E. Smith, *The Journals* and G. L. Smith, *The Journals*.

25. Daniel Pratt, "Cotton Gins," *De Bow's Review* 2 (September 1846): 153.

26. Ibid.

27. F. E. Smith, *The Journals*, 4–5, 44, 56–57, 67, 78, 80, 86, 105, 109.

28. Ibid., 87.

29. Ibid., 106.

30. G. L. Smith, *The Journals*, 7.

31. F. E. Smith, *The Journals*, 89.

32. Ibid., 93.

33. G. L. Smith, *The Journals*, 9.

34. Ibid., 19–20.

35. Ibid., 21–22.

36. F. E. Smith, *The Journals*, 106.

37. Ibid., 115.

38. G. L. Smith, *The Journals*, 103.

39. Daniel Pratt, "Alabama Improvements and the True Interests of her People," *American Cotton Planter and Soil of the South* n.s. 3 (April 1859): 114–15.

40. Lyddleton Smith, Alabama, vol. 11, p. 190, R. G. Dun & Co.

41. Du Bois is discussed in chapter 6.

42. Evans, *Conquest of Labor*, 241.

43. *Attala County Mississippi Cemeteries* (Jackson, Miss.: Miss. Dept. Archives and History, n.d.), 123.

44. Kosciusko-Attala Historical Society, *Kosciusko-Attala History* (s.l.: s.n., 1976), 55, 159; Ruby Haynes, "David Chase Atwood, Jr.," Miss. Dept. of Archives and History, typescript. Because there is so little known about T. G. Atwood, the Haynes source is important but it contains numerous factual errors.

45. *[Aberdeen, Miss.] Weekly Independent,* 30 December 1848; *[Vicksburg] Whig,* 6 November 1858, 19 October 1859; Mississippi, vol. 21, p. 24, R. G. Dun & Co.

46. Mississippi, vol. 21, p. 24, R. G. Dun & Co.

47. M. W. Philips, "Gin Stands," *Southern Cultivator* 8 (1850): 56–57.

48. *[Vicksburg, Miss.] Sentinel,* 16 April 1850, in Moore, *Emergence of the Cotton Kingdom,* 218.

49. *[Vicksburg, Miss.] Whig,* 6 November 1858, in Moore, *Emergence of the Cotton Kingdom,* 218.

50. Amasa Davis, "Gin Stand Manufactory," *Mississippi Free Trader and Natchez Gazette,* 11 March 1847.

51. James H. Stone, "The Economic Development of Holly Springs during the 1840s," *Journal of Mississippi History* 32 (1970): 357. Stone documented the Holly Springs Foundry as the only heavy industry in the city but located a range of other manufacturing establishments.

52. *Columbus [Miss.] Democrat,* 28 January 1837 (Aikin), 22 April 1837 (Aikin), 15 April 1837 (Jones), 6 January 1838 (Collier), 20 April 1839 (McIntyre).

53. Mississippi, vol. 14, p. 26, R. G. Dun & Co.; *Columbus [Miss.] Democrat,* 7 November 1840.

54. Mississippi, vol. 14, p. 24. R. G. Dun & Co.

55. S. S. Ward, "To the Farmer," *Aberdeen [Miss.] Advertiser,* 4 January 1845.

56. B. D. Gullett is discussed in chapter 6.

57. Wm. T. McCracken, "Workmen Wanted Immediately," *Aberdeen [Miss.] Weekly Independent,* 24 May 1851.

58. Moore, *Emergence of the Cotton Kingdom,* 168–69.

59. Beach & Keller, "Gin Wrights," *Woodville [Miss.] Republican,* 10 July 1841 (dated 13 March 1841).

60. V. N. H. Netterville, *Woodville [Miss.] Republican* 10 July 1841, 1 January 1848.

61. P. B. Tyler to J. Alexander Ventress, 13 October 1856, Joseph E. Carver to J. Alexander Ventress, 22 March 1858, J. A. Tuttle to J. Alex. Ventress, 4 November 1857, 8 May 1858, Trask-Ventress Family Papers, Miss. Dept. of Archives and History, Jackson, Miss.; J. A. Ventress, of Woodville, Miss., "Improvement in Cotton-Gins, U.S. Pat. No. 20,717 (1858). Cavett Taff graciously brought these papers to my attention and made them available.

62. Samuel Griswold to Daniel Pratt, Clinton, [Ga.], 29 January 1850, Pratt Family Papers, Alabama Department of Archives and History, Montgomery (hereafter ADAH).

63. R. Abbey, "Cotton and the Cotton Planter," *De Bow's Review* 3 (January 1847): 11.

64. *Augusta Directory and City Advertiser* (Augusta, Ga.: Browne & McCafferty, 1841), 58.

65. Thomas J. Cheely, "Cotton Gin Factory," *Southern Business Directory and General Commercial Advertiser* (Charleston, S.C.: Press of Walker & James, 1854), 288.

66. Richard T. Hill and William E. Anthony, *Confederate Longarms and Pistols* (Charlotte, N.C.: The Authors, 1978), 60–63.

67. Louisiana, vol. 11, p. 355, R. G. Dun & Co. Gunnison was also a partner in Gunnison, Chapman & Co., a manufacturer of Benjamin D. Gullett's cotton gins in New Orleans.

68. "Manufacture of Colt's Revolver," *Macon [Ga.] Telegraph,* 5 August 1862.

69. W. C. Hodgkins to R. M. Cuyler, 16 July 1862, Letters received by Capt. Richard M. Cuyler, June-July 1862, 211, in Spengler, *Griswoldville,* 77.

70. R. Milton Cary to Richard M. Cuyler, 22 October 1862, Letters received by Major Richard M. Cuyler, September-October 1862, chap. 4, vol. 6, p. 218, in Spengler, *Griswoldville,* 78.

71. James H. Burton to Colonel [R. M. Cuyler?], 1 November 1864, Confederate Papers Relating to Citizens or Business Firms, Roll 383, in Spengler, *Griswoldville,* 79–80.

72. Samuel Griswold to James H. Burton, 28 October 1864, Spengler, *Griswoldville,* 79–80.

73. Hill and Anthony, *Confederate Longarms and Pistols,* 265–66; William B. Edwards, *Civil War Guns* (Harrisburg, Pa.: Stackpole Co., 1962), 354.

CHAPTER SIX: Saw Gin Innovation, 1820–1860

1. Gray, *History of Agriculture,* 1026–27: tables 40–41.

2. Elisha Reid, "Saw-Gin Business," *The [Milledgeville, Ga.] Reflector,* 16 June 1818.

3. William A. Aikin, "Cast Steel and Carver Saw Gins," *Florence [Ala.] Gazette,* 19 August 1824.

4. Eleazer Carver, Petition for Extension, Filed 4 October 1858 on U.S. Pat. No. 3,875 (1845), Box 51, Extension Case Files, 1836–75, Records of the Patent Office, RG 241, NARA. See also E. Carver, "To the Editors," *American Polytechnic Journal* 2 (July-December 1853): 383.

5. J. R. Bedford, "Improved Mode of cleaning Cotton," *American Farmer* 3 (2 November 1821): 256.

6. Eleazer Carver, Petition for Extension.

7. Ibid.

8. Ross Thomson, "Crossover Inventors and Technological Linkages: American Shoemaking and the Broader Economy, 1848–1901," *Technology and Culture* 32 (October 1991): 1018, 1025. The theory on crossovers and convergence that Thomson develops for the shoemaking industry applies to the gin-making industry.

9. R. Collins, "Management of Cotton," *Southern Agriculturist* n.s. 2 (November 1842): 566–67. Reprinted from *Southwestern Farmer.*

10. R. Abbey, *De Bow's Review* 2 (September 1846): 137–38.

11. Eleazer Carver, U.S. Pat. 777 (1838); Reissue No. 17 (1839).

12. *Eleazer Carver* v. *Joseph A. Hyde et al.,* United States Supreme Court Reports, 10 Law. Ed., 16 Pet. (41 U.S.) 515, 519.

13. Bates, Hyde & Co., "To the Cotton Planters," 10 August 1843. Owned privately but reproduced in "Improved Eagle Cotton Gin," *American Agriculturist* (1848): 152.

14. *Eleazer Carver v. Braintree Manuf'g Co.,* 5 *Fed. Cas.,* Case No. 2,485, pp. 236, 238, 240, 242.

15. Ibid.

16. John Du Bois, U.S. Pat. No. 6,998 (1850); C. A. McPhetridge, U.S. Pat. No. 2,068 (1841); James F. Orr, U.S. Pat. No. 19,041 (1858).

17. Wm. K. Orr, "Gins—An Important Discovery," *American Cotton Planter & Soil of the South* 2 (May 1858): 152.

18. W. B. Stewart, U.S. Pat. No. 3,097 (1843); John Simpson U.S. Pat. No. 13,441 (1855); Alexander D. Brown U.S. Pat. No. 13,484 (1855).

19. Wailes, *Report,* 171.

20. Robert W. Smith & Co., "Carver's Improved Cotton Gins," *Montgomery [Ala.] Daily Mail,* 20 November 1854.

21. *"The Taylor" Cotton Gin* (Columbus, Ga.: W. G. Clemons, Brown & Co., [1870]), 3–4.

22. Peter Von Schmidt, U.S. Pat. No. 4,817 (1846).

23. Ibid.

24. Robert A. L. McCurdy, U.S. Pat. 13,131 (1855); Federal Census 1860, La., Natchitoches Co.

25. U.S. Pat. No. 13,641 (1855); "Fultz Patent Cotton Gin," *Scientific American* 11 (3 November 1855): 64.

26. William F. Pratt, assignor to the E. Carver Company, U.S. Pat. No. 25,307 (1859). Pratt was not related to Daniel Pratt.

27. E. Carver Company, *Patent Roll-Clearer for Cotton Gins* (East Bridgewater, Mass.: E. Carver Company, [1859]), [2]. Original flyer at Museum of American Textile History, Lowell, Mass.

28. William F. Pratt, assignor to the E. Carver Company, U.S. Pat. No. 25,307 (1859).

29. David G. Olmstead, U.S. Pat. No. 19,097 (1858).

30. Theodorick J. James, U.S. Pat. No. 2,608 (1842).

31. A, "Improved Plans of Preparing and Ginning Cotton," *Southern Agriculturist* n.s. 3 (October 1843): 386–87.

32. Ibid.

33. *Southern Agriculturist* n.s. 3 (December 1843): 467–68.

34. "The Cotton-Gin," *American Polytechnic Journal* 2 (July to December 1853): 262.

35. G. McFarlane & Co. et al., "Cotton Gin Notice," *[Aberdeen, Miss.] Weekly Independent,* 1 December 1849.

36. Compare Carver's guard reprinted in *American Polytechnic Journal* 2 (July to December 1853): 262, with Keith's reprinted in *Appleton's Dictionary of Machines, Mechanics, Engine-Work, and Engineering* (1852), s.v. "Gin," plate 2121.

37. Clay Lancaster, *Eutaw: The Builders and Architecture of an Ante-bellum Southern Town* (Eutaw, Ala.: Greene County Historical Society, 1979), 48–49.

38. R. C. Beckett, "Antebellum Times in Monroe County," *Publications of the Mississippi Historical Society* 11 (1910): 89. Beckett identified Richard Gladney as a "cousin" of Gullett.

39. Gullett Gin Company, *A Century of Progress, 1849–1949* (Amite, La.: Gullett Gin Company, [1949], 1. Original in the Warshaw Collection of Business Americana, Ar-

chives Center, National Museum of American History, Smithsonian Institution, Washington, D.C.

40. Lancaster, *Eutaw,* 79.

41. Ibid.; *Columbus [Miss.] Democrat,* 7 November 1840.

42. Mississippi, vol. 14, p. 26, R. G. Dun & Co.

43. *Scientific American* 9 (1 October 1853): 20; Leonard Campbell, U.S. Pat. No. 10,401 (1854), No. 12,894 (1855).

44. Beckett, "Antebellum Times in Monroe County," 89.

45. Noah B. Cloud, *American Cotton Planter and Soil of the South,* n.s., 11 (June 1858): 168.

46. M. W. Philips, "Cotton Gins and Gin Feeders," *Southern Cultivator,* 18 (November 1860): 339.

47. G. D. Harmon, "Gullett's Cotton Gin," *Southern Cultivator,* 18 (November 1860): 343. Italics in original.

48. "The Great Southern Gin, or Patent Steel Comb and Brush Gin," *Daily Vicksburg [Miss.] Whig,* 24 November 1860.

49. Benjamin D. Gullett, U.S. Pat. No. 140,365 (1873).

50. A. Q. Withers, U.S. Pat. No. 21,714 (1858)

51. Evans, *Conquest of Labor,* 42–43.

52. M. W. Philips, *Southern Cultivator* 18 (November 1860): 339.

53. Ibid.

CHAPTER SEVEN: Old and New Roller Gins, 1820–1870

1. Thomas Spalding, "Cotton—Its Introduction and Progress of Its Culture, in the United States," *Southern Agriculturist* 8 (January 1835): 46.

2. Thomas Spalding, "Cotton," *Southern Agriculturist* 8 (February 1835): 82–83.

3. Ibid.

4. In 1819, Upland cotton sold for 10 pence per pound and Sea Island cotton for 21 pence per pound. In the years that followed, the corresponding figures are: 1828 (5, 10), 1837 (7, 16), 1840 (5.75, 13.50), and 1845 (3.38, 10.20). Cotton prices at Liverpool, England, from Richard Burns, *Statistics of the Cotton Trade* (London: Simpkin, Marshall & Co., 1847), 22. U.S. customs officials did not disaggregate Sea Island cotton from short-staple cotton varieties before 1802 and thereafter kept less complete export records than did British officials.

5. Some long-staple cotton growers also grew short-staple cotton. See Malcolm Bell Jr., *Major Butler's Legacy: Five Generations of a Slaveholding Family* (Athens: University of Georgia Press, 1987), 112; Thomas Walter Peyre, Plantation Journal, South Carolina Historical Society, Charleston, S.C., text-fiche, passim; John D. Campbell, "The Gender Division of Labor, Slave Reproduction and the Slave Family Economy on Southern Cotton Plantations, 1800–1865" (Ph.D. diss., University of Minnesota, 1988), table 1–1.

6. *Southern Agriculturist* 1 (1828): 17, 26.

7. William Elliot, "On the Cultivation and High Prices of Sea-Island Cotton," *Southern Agriculturist* 1 (April 1828): 154.

8. Spalding, "Cotton," *Southern Agriculturist* 8 (January 1835): 44.

9. James Harvie, "Specification of the Patent," [London, England] *Repertory of Arts, Manufactures, and Agriculture,* second series 45 (1824): 14–15. Drawings were reproduced in plate II.

10. Whitemarsh B. Seabrook, *Memoir on the Origin, Cultivation and Uses of Cotton* (Charleston, S.C.: Printed by Miller & Browne, 1844), 33. Seabrook was governor of South Carolina from 1848–50.

11. J. B. W., "Whittemore's Cotton Gin," *Southern Agriculturist* 8 (September 1835): 475.

12. Amos Whittemore, "Machine for Manufacturing Sheet Cards," U.S. Pat. Unnumbered (1797); William R. Bagnall, *The Textile Industries of the United States* (Cambridge, Mass., 1893; reprint, New York: Augustus M. Kelley, 1971), 155; Lucius R. Paige, *History of Cambridge, Massachusetts, 1630–1877 with a Genealogical Register* (New York: Hurd and Houghton, 1877; reprint, Bowie, Md.: Heritage Books, 1986), 690.

13. J. B. W., "Whittemore's Improved Cotton Gin," 333.

14. Ibid., 473–77.

15. Ibid., 479.

16. Seabrook, *Memoir on Cotton,* 34.

17. John D. Legare, "Account of an Agricultural Excursion made into the South of Georgia in the winter of 1832; by the Editor," *Southern Agriculturist* 6 (April 1833): 161, 244.

18. Ibid., 245.

19. Ibid., 244–45.

20. Ibid., 245.

21. George Jones Kollock, Plantation Records, 1837–61, 10 November 1837, Southern Historical Collection, University of North Carolina at Chapel Hill, microfilm.

22. J. Cart Glover, "Pottle's Improved Cotton-Gin," *Southern Agriculturist* 5 (March 1845): 117; Seabrook, *Memoir,* 34.

23. Glover, "Pottle's Improved Cotton-Gin," 116.

24. R. Furman, "Inquiries Retative [*sic*] to the Best Gins, and the Proper Mode of Packing Cotton for Market," *Southern Agriculturist* n.s. 3 (October 1843): 393.

25. Ibid., 394–95.

26. Fones McCarthy, U.S. Pat. No. 1,675 (1840), handwritten copy, RG 241, Box No. 24, Patent Office Files, NARA, Suitland, Md.; Fones McCarthy, "Reasons Given," Patent Extension Case Files, 1836–75, Box 201, RG 241, E18, NARA, hereafter, Patent Extension File.

27. Argument of P. H. Watson, [14], Patent Extension File. Underlined in original.

28. Fones McCarthy, U.S. Pat. No. 1,675 (1840); Reissue No. 262 (1854).

29. Ibid.

30. Argument of P. H. Watson, [2–3], Patent Extension File.

31. Seabrook, *Memoir on Cotton,* 35.

32. Whitemarsh B. Seabrook, "From the Charleston Mercury. Mr. McCarthy's Cotton Gin, " *Southern Agriculturist* n.s. 4 (May, 1844): 183. Italics in original.

33. Ibid., 185.

34. McCarthy, "Reasons Given," [3], Patent Extension File.

35. Ibid., 3–4.

36. The foundry operated through the decade and advertised frequently. For example see *[Jacksonville] Florida News,* 26 March 1853.

37. McCarthy, "Reasons Given," [6], Patent Extension File.

38. Ibid.

39. Ibid.; "Statement of Receipts and Expenditures," [5], Patent Extension File.

40. Robert F. W. Allston, "Sea-Coast Crops of the South," *De Bow's Review* 16 (June 1854): 597–98.

41. Ibid.

42. "We the undersigned," 6 June 1854, McCarthy Patent Extension File.

43. W. S. Henery & Co., *The Charleston Directory* (Charleston, S.C.: Walker, Evans & Co., 1859), 27; *Directory of the City of Charleston* (Savannah, Ga.: John M. Cooper & Co., 1860), 70. Henery advertised himself as a "manufacturer of steam engines, boilers and machinery" and may also have made the McCarthy gin. He also carried (or made) the Osgood gin, patented by Enoch Osgood of Boston, Massachusetts. It operated on a similar principle to McCarthy's.

44. Report of the Examiner, 17 June 1854, McCarthy Patent Extension File. McCarthy also received a reissue (no. 262) on the 1840 patent on 18 April 1854.

45. Editor, "M'Carthy's Cotton Gin," *Southern Agriculturist* n.s. 5 (January 1845): 27.

46. Leonard Wray, "The Culture and Preparation of Cotton in the United States of America," *Journal of the Royal Society of Arts* 7 (December 1858): 82.

47. Ibid.

48. Israel F. Brown, U.S. Pat 39,767 (1863); U.S. Pat No. 45,695 (1865). Brown's 1865 patent was for a roller for the McCarthy gin.

49. Milton Whipple, Patent Extension Case Files, 1836–71, Box No. 292, RG 241, NARA.

50. Milton D. Whipple, U.S. Pat. No. 1,839 (1840).

51. Lewis G. Sturdevant, U.S. Pat. No. 2,190 (1841); Francis A. Calvert, U.S. Pat. No. 2,373 (1841); Francis A. Calvert U.S. Pat. No. 3,120 (1843), Reissued 27 December 1843, No. 59; Theodore Ely, U.S. Pat. No. 3,269 (1843).

52. Stephen R. Parkhurst, U.S. Pat. No. 4,023 (1845); Extension, 26 April 1859; Reissue No. 1,137 (1861).

53. "Parkhurst's Cotton Gin," *Scientific American* 5 (12 January 1850): 130.

54. T[homas] A[ffleck], "Stock of Dr. Davis-New Cotton Gin-McComb's Press," *Southern Cultivator* o.s. 9 (February 1851): 24.

55. *Mobile Daily Advertiser,* 24 August 1852, reprinted from the *Savannah Republican; Mobile Daily Advertiser,* 7 November 1852, reprinted from the *Savannah Courier,* both in Curtis J. Evans, Daniel Pratt: Yankee Industrialist in the Antebellum South (Master's Thesis, Louisiana State University and Agricultural and Mechanical College, 1993), 84.

56. F. E. Smith, *The Journals,* 115.

57. Ibid., 117.

58. Ibid., 126–27, 129.

59. Ibid., 155, 157–58, 148, 169.

60. Samuel Griswold to Daniel Pratt, 29 January 1850, Samuel Griswold to Daniel Pratt, 7 June 1852, Pratt Family Papers, ADAH.

61. Daniel Pratt, U.S. Pat. No. 17,806 (1857).

62. Daniel Pratt & Co., "Cotton Gin Manufactory," *Autauga [Ala.] Citizen,* 3 February 1853; *[Aberdeen, Miss.] Weekly Independent,* 16 April 1853.

63. Stephen R. Parkhurst, U.S. Pat. No. 20,086 (1858), Extended 1872.

64. Edwin T. Freedley, *Philadelphia and Its Manufactures* (Philadelphia: Edward Young, 1858), 301; Thomas R. Navin, *The Whitin Machine Works Since 1831: A Textile Machinery Company in an Industrial Village* (Cambridge: Harvard University Press, 1950), 206.

65. Alfred Jenks & Son, "To Cotton Planters," *De Bow's Review* 25 (October 1858): n.p.

66. Planters corroborated this claim. They wrote that they ginned wet and "rotten" cotton on the gin. See H., "Cotton Gins-Cotton Seed for Hogs," *Southern Cultivator* 17 (December 1859): 359.

67. *De Bow's Review* o.s. 26 (February 1859): 238.

68. Martin W. Philips, "Saw Gins, &c." *Southern Cultivator* 17 (August 1859): 238.

69. Martin W. Philips, *Southern Cultivator* (1856): 366.

70. William K. Orr, "Cotton Gins and Their Management," *American Cotton Planter and Soil of the South* (1857): 104.

71. *Whipple* v. *Baldwin Manufacturing Company*, Circuit Court, Mass., 29 *Fed. Cas.* (1858), 931–33.

72. *Stephen R. Parkhurst* v. *Israel Kinsman and Calvin L. Goddard*, Circuit Court, N.Y., 18 *Federal Cases* (1849), 1200, 1205; *Abstract of Case, Pleadings & Evidence on the Part of the Appellee [Parkhurst], Israel Kinsman and Calvin L. Goddard* v. *Stephen R. Parkhurst, United States Supreme Court* (New York: Wm. C. Bryant, 1856). Parkhurst included this document as Exhibit G in his Patent Extension File.

CHAPTER EIGHT: Machine and Myth

1. R[ichard] L. Allen, *The American Farm Book* (New York: A. O. Moore, 1858), 198–99. The work was copyrighted in 1849 and also published in New York by C. M. Saxton and Company in 1857.

2. Olmsted, "Memoir of Eli Whitney," 214.

3. Dexter, *Biographical Sketches*, s.v. "Denison Olmsted," 5:592–600; Denison Olmsted, U.S. Patent Restored, Not Numbered, reel 1 (1826–27).

4. Olmsted, "Memoir of Eli Whitney," 214.

5. Ibid., 211, 216–17.

6. Ibid., 208.

7. Ibid., 209.

8. S., "Sketch of the Life of the Late Eli Whitney, with Some Remarks on the Invention of the Saw-Gin," *Southern Agriculturist* 5 (August 1832): 393–403. Scholars agree that "S." was South Carolina–born William Scarbrough or Scarborough who financed the *Savannah*, a steamship that made a pioneer transatlantic voyage. It left the Georgia coast on 24 May 1819 and arrived at Liverpool on 20 June and St. Petersburg on 13 September before returning to Savannah on 30 November. Italics in original. See also Frederick L. Lewton, *Historical Notes on the Cotton Gin* (Washington, D.C., 1938): 554, 557.

9. Edward Baines, *History of the Cotton Manufacture in Great Britain* (London: H. Fisher, P. Fisher, and P. Jackson, 1835), 299–300. Baines was the antilabor, antislavery editor of the *Leeds Mercury* and used the *History* as a platform to criticize opponents of child labor.

10. George S. White, *Memoir of Samuel Slater, The Father of American Manufactures* (Philadelphia: Printed at No. 46 Carpenter Street, 1836): 347–49, 351, 368–75.

11. B. L. C. Wailes, *Address to the [Washington, Miss.] Agricultural, Horticultural, and Botanical Society of Jefferson College* (Natchez: Printed at the Daily Courier Office, 1841), 12.

12. Seabrook, *Memoir on Cotton*, 11.

13. *Appleton's Dictionary of Machines, Mechanics, Engine-Work, and Engineering* (New York: D. Appleton & Company, 1852), 1:862.

14. Benjamin Silliman Jr., and C. R. Goodrich, eds., *The World of Science, Art, and Industry Illustrated* (New York: G. P. Putnam and Company, 1854), n.p.

15. William Barbee, *The Cotton Question: The Production, Export, Manufacture, and Consumption of Cotton* (New York: Metropolitan Record Office, 1866), 94.

16. Compare "Ginning by Hand in 1792," *Cotton Seed Oil Magazine*, 1 February 1916 to "Sorting Cotton," *Frank Leslie's Illustrated Newspaper* 17 April 1869, and to an 1862 photograph of women either moting or sorting and preparing seed cotton for the gin taken by T. H. O'Sullivan. Prints & Photographs Div., Library of Congress, negative number LC-B8171-159.

17. "The First Cotton-Gin," *Harper's Weekly* 13 (18 December 1869): 814; engraving, 813.

18. The analysis of the image and idea of Aunt Jemima presented in M. M. Manring, *Slave in a Box: The Strange Career of Aunt Jemima* (Charlottesville and London: University Press of Virginia, 1998) was helpful in reading this and other postbellum illustrations of African Americans at work.

19. Michel Pastoureau, trans. Jody Gladding, *A History of Stripes and Striped Fabric* (New York: Columbia University Press, 2001). Pastoureau explains that in the Middle Ages, stripes signified depravity. Their meaning changed when they crossed the Atlantic, but old associations persisted in the use of striped fabric for prison inmate uniforms.

20. S., "Sketch of the Life of the Late Eli Whitney," 395, 397.

21. A Subscriber, "Anecdote Relative to the Origin of Whitney's Saw-Gin," *Southern Agriculturist* 5 (September 1832): 469. Italics in original.

22. A Small Planter, "Remarks on the 'Sketch of the Life of Eli Whitney,' &c." *Southern Agriculturist* 5 (December 1832): 628–29.

23. A Small Planter, "Further Remarks on the 'Life of Eli Whitney,'" *Southern Agriculturist* 6 (February 1833): 80.

24. Joseph A. Turner, *The Cotton Planter's Manual* (New York: C. M. Saxton & Co., 1857; reprint New York: Negro University Press, 1969), 303.

25. Antiquary, "Cotton Gin and Packing Screw;" C., "Nathan Lyons;" Thomas H. White, "Origin of the Cotton Gin,"in Turner, *Manual*, 289, 293–94, 289–90.

26. Rev. Henry Ward Beecher, *Harper's Weekly* (14 November 1863): 733.

27. C. W. Grandy, *The Cotton Question* (Norfolk, Va.: C. W. Grandy & Sons, 1876), 4–5.

28. John L. Hayes, *American Textile Machinery* (Cambridge: Cambridge University Press, 1879), 36.

29. Edward C. Bates, "Eli Whitney: The Story of the Cotton-Gin," *New England Magazine*, n.s. 2 (May 1890): 288.

30. Alice Hyneman Rhine, "Woman in Industry," in Annie N. Meyer, ed., *Woman's Work in America* (1891; reprint, New York: Arno Press, 1972), 279–80; Stegeman and Stegeman, *Caty*, 2, 19. The Stegemans based the biography on primary evidence revealing that Greene abhorred handwork, even "stitchery."

31. Rhine, "Woman in Industry," 279–80.

32. James Ford Rhodes, *History of the United States from the Compromise of 1850* (London: Macmillan and Co., 1893), 19, 25, 26–27.

33. Hammond, "Correspondence," 90–91. See also his book *The Cotton Industry: An Essay in American Economic History* (New York: Macmillan, 1897), 22–31.

34. Hammond, "Correspondence," 97–98.

35. *Cotton Ginning Machinery for Separating any description of Cotton from its Seed, Exhibited by Platt Bros. & Co. Limited, Hartford Works, Oldham, England* (Philadelphia: Centennial Exhibition, 1876), 2. Original in Box 1, Cotton, Collection No. 60, Warshaw Collection of Business Americana, Archives Center, National Museum of American History, Smithsonian Institution, Washington, D.C.

36. J[ohn] Forbes Watson, *Report on Cotton Gins and on the Cleaning and Quality of Indian Cotton* (London: William H. Allen & Co., 1879), 14–15. In other categories, Watson ranked the Emery Gin, manufactured in Albany, New York, superior to the Gullett gin.

37. Henry W. Grady, "Cotton and Its Kingdom," *Harper's New Monthly Magazine* 63 (1881): 729.

38. Edward Atkinson, "Improvement of Cotton," in *Transactions of the New England Cotton Manufacturers Association* (hereafter *NECMA*) 59 (24–25 October 1895): 111, 120, 122, 125; Edward Atkinson quoted in Charles J. McPherson, "Improved Ginning," in *Transactions of the NECMA* 74 (22–23 April 1903): 220–21.

39. "Ginning by Hand in 1792," *Cotton Seed Oil Magazine,* 1 February 1916; "Hand Cleaning Cotton," in *Patriots Plant Cotton in 1776* (Libertyville, Ill.: Rainbow Division, International Minerals & Chemical Corp, 1975). The 1916 image graphically interprets Olmsted, picturing a man of African descent dozing while plucking cotton fiber from seed. The 1975 image was reprinted from an earlier, unidentified source. It pictures a black man seated on a stool, leaning against his cabin, dozing as he plucks fiber from the seed.

40. Hodding Carter, Review of David L. Cohn, *Life and Times of King Cotton, New York Times,* 11 November 1956.

41. David L. Cohn, *The Life and Times of King Cotton* (New York: Oxford University Press, 1956), 3–4.

42. Stampp, *The Peculiar Institution,* 5, 7, 46, 66. See Fogel and Engerman (1974) who cited Stampp as perpetuating the "Myth of Black Incompetence." Stampp saw as his goal not overturning Olmsted but the Ulrich B. Phillips's school that argued for the inherent inferiority of Africans and the ameliorative benefit of slavery.

43. Melissa Glass, "The Touch . . . the Feel . . . of Eli Whitney," *Opelika-Auburn [Ala.] News,* 20 August 2001.

Essay on Sources

Bits of evidence gleaned from archives, city directories, newspapers, census manuscripts, patents, litigation, and R. G. Dun & Company credit reports make up American cotton gin history. Of these, census manuscripts, newspapers, and patent documents provide the most useful information, particularly about the ordinary gin maker. For individuals who left no personal or business records, the census manuscripts are a critical source of information. They give occupational title and wealth, places of birth for spouses and children, along with agricultural and manufacturing activity. The slave schedules list the enslaved by owner, then by age and physical characteristics, excepting the rare instance when a name or occupation is noted. The National Archives and Records Administration (NARA) in Washington, D.C., and in regional offices, makes them available on microfilm but they can be purchased through—and used at—the Wilson Library of the University of North Carolina.

Newspaper notices identify many gin makers missed by census marshals; like city directory listings, they add context and complexity. For that, the *Augusta [Ga.] Chronicle,* begun in 1785 and preserved in a complete run by the University of Georgia Library, was most useful, as was the *[Savannah] Georgia Gazette,* begun in 1763. Apart from university libraries and state archives and history departments, the Library of Congress in Washington, D.C., is the most important repository of early newspapers. They have preserved discontinuous "portfolio" issues of local newspapers like *The [Milledgeville, Ga.] Reflector* through the 1810s, the *[Tuscumbia, Ala.] Franklin Enquirer* through the early 1820s, and the *[Natchez] Mississippi State Gazette* from the 1810s through the 1820s, among literally hundreds of others, along with runs of journals like the *Southern Agriculturist, American Farmer,* and *De Bow's Review.* These are invaluable sources of information about machine making in the early national and antebellum South that must be read slowly and carefully, lest the eye gloss over the single-line notice placed by a parsimonious mechanic. The able staff of the Museum of Early Southern Decorative Arts in Winston-Salem, North

Carolina, offers an alternative. They have scoured newspapers, city directories, and government documents including the manufacturing census of 1820 and organized relevant citations into a database of craftsmen covering the colonial period through 1820. Information is accessible by name or area, date, trade, and type of source. Historic Brattonsville in McConnells, South Carolina, has a similar database of antebellum South Carolina artisans.

Patent documents include the descriptions and drawings as well as reissued patents and extension files. Carolyn Cooper, Steve Lubar, Kendall J. Dodd, and Ross Thomson in articles in a special issue on "Patents and Invention" in *Technology and Civilization* 32 (October 1991), examined the United States patent system, its origins and function, how patentees used it, and what it meant in an age that scorned monopoly. These articles sent me to patent extension files, which proved the most valuable source in this area. The files include biographies of the applicants, business histories, broadsides and trade catalogs, along with the petition, substantiating evidence and counterevidence. Had Fones McCarthy not filed his last-minute reissue and extension, virtually nothing would be known of him. Benjamin D. Gullett's file included trade catalogs that linked him with Mystic River Hardware, where McCarthy had his gins built. They allowed other important linkages to be made between gin makers and mechanics.

For wealthy gin makers like Joseph Eve, Eli Whitney, and Daniel Pratt, there are ample manuscript sources. Sources on Eve are scattered in Philadelphia, the Bahamas, and Augusta, Georgia; the most valuable on Eve himself—as opposed to his family—are in the Bahamas. The Bahamas Department of Archives is a division of the British Public Records Office. In addition to unpublished and published manuscript materials, it holds a run of the *Bahama Gazette,* where Eve publicized his gins and also served as editor. Yale University's Sterling Library holds the Eli Whitney Papers and the Blake Family Papers, which include correspondence between Whitney and Phineas Miller, Miller's two letterbooks, and records for his eight ginneries. The library also owns an original copy of Joseph Eve's poetic opus *Better to Be.* Daniel Pratt's letters and business records are at the University of North Carolina, Chapel Hill; Auburn University, Alabama; the Alabama Department of Archives and History, Montgomery; as well as the corporate offices of Continental Eagle in Prattville, Alabama, where Tommy M. Brown ably curates them. The Autauga County Heritage Association owns the Smith Brothers diaries, which are an unparalleled source of information on daily shop practice.

For secondary sources on gins and ginning, the literature published by the United States Department of Agriculture is unequaled. They are generally directed at the nonspecialist and typically include a historical component. The department regularly updates its *Cotton Ginners Handbook* (ARS Agricultural Handbook No. 503). Following the texts over time is itself instructive of gin and industry change. The 1994 edition opens with a brief history of the gin by William D. Mayfield and W. S. Anthony, and includes articles on cotton production, harvesting, and storage. Articles on ginning—particularly S. Ed Hughs et al., on "Moisture Control"; R. V. Baker et al., on "Seed Cotton Cleaning and Extracting"; G. J. Mangialardi Jr. et al., on "Lint Cleaning"; and Jesse F. Moore on "The Classification of Cotton"—cover contemporary issues that illuminate past practices.

There are no historical studies on the cotton gin. As a machine, however, it is incorporated conceptually within the literature of southern industrialization. Gins and gin makers are referenced specifically in *A Deplorable Scarcity: The Failure of Industrialization in the Slave Economy,* Fred Bateman and Thomas Weiss's study of antebellum southern industrialization (University of North Carolina Press, 1981). In this essential text, the authors reviewed the historiography by addressing the "major hypotheses" of assumed "backwardness," testing the most plausible set against the evidence of the census data of 1850 and 1860. They concluded that planter behavior—as opposed to market size, returns on manufacturing investment, or the slave labor force—slowed the pace of industrialization. Their methodology of census data correlation influenced the simpler application here. The authors' use of only the manufacturing census data in their industry profiles inspired the search through population schedules for evidence of excluded gin makers. Although it added dimension to southern industry, the additional data would not have altered the conclusion. In addition to describing their sample set, the authors discussed the problems of data gathering and census use (Table A-2, pp. 168–71).

Late twentieth-century historians have built on Bateman and Weiss's assumptions, examining planter behavior as it impinged on the paradoxes of southern industry. Claudia D. Goldin reconciled the industrializing, modernizing, and slave-labor–based South in her 1976 *Urban Slavery in the American South, 1820–1860: A Quantitative History* (University of Chicago Press), showing how movement of the enslaved between town and country reflected elasticity of demand but also occupational flexibility. In *An Anxious Pursuit: Agricultural Innovation and Modernity in the Lower South, 1730–1815* (University of

North Carolina Press, 1993), Joyce Chaplin found fraught planters modernizing as they reinforced medieval social structures. Ernest McP. Lander argued in
his 1969 *The Textile Industry of Antebellum South Carolina* (Lousiana State University Press) that the War of 1812 provided an industrializing impulse. For
John Hebron Moore in his *Emergence of the Cotton Kingdom in the Old Southwest, Mississippi, 1770–1860* (Lousiana State University Press, 1988), the panic
of 1837 propelled modernization and industrialization; momentum sustained
them. Curtis J. Evans, in his biography of Daniel Pratt, *The Conquest of Labor:
Daniel Pratt and Southern Industrialization* (Lousiana State University Press,
2001), analyzed a southerner by residence who disregarded planter hegemony, both manufacturing and planting on a large scale in central Alabama.

Charles B. Dew in *Ironmaker to the Confederacy: Joseph R. Anderson and the
Tredegar Iron Works* (1966; Library of Virginia, 1999) and *Bond of Iron: Master
and Slave at Buffalo Forge* (Norton, 1994) wrote of iron forgers and founders,
not machine makers specifically, yet he examined the dense network of reciprocal social relations that moderated labor exchange and communication between master and slave in those and other machine skill–based industries.
Dew's protagonists Joseph Anderson at Tredegar and William Weaver at Buffalo Forge both managed racially integrated labor forces, as did gin makers like
Daniel Pratt; they also motivated the enslaved artisans with overwork, as Pratt
did. Dew tapped a deep archive of business records and correspondence to
bring owners and the enslaved ironworkers and their families to life. White
and black, they shifted in and out of industrial and agricultural pursuits as
they experienced joys and tragedies, including slave sales, reconciling the par-
adox of modernity and slavery by living it.

Index

Lightning Source UK Ltd.
Milton Keynes UK
UKHW010626160322
400106UK00001B/21

9 780801 882722